Farmstead and Artisan Cheeses

A GUIDE TO BUILDING A BUSINESS

Barbara Reed
Dairy Farm Advisor (retired)
University of California Cooperative Extension
Central Valley Region

Leslie J. Butler
University of California Cooperative Extension Specialist
Department of Agricultural and Resource Economics
University of California, Davis

Ellie Rilla
Community Development Advisor
University of California Cooperative Extension
Marin County

■ **University** *of* **California** Agriculture and Natural Resources
Publication 3522

M000291220

To order or obtain ANR publications and other products, visit the ANR Communication Services online catalog at http://anrcatalog.ucdavis.edu or phone 1-800-994-8849. You can also place orders by mail or FAX, or request a printed catalog of our products from

University of California
Agriculture and Natural Resources
Communication Services
1301 S. 46th Street
Building 478 - MC 3580
Richmond, CA 94804-4600

Telephone 1-800-994-8849
510-665-2195
FAX 510-665-3427
E-mail: anrcatalog@ucdavis.edu

Publication 3522
ISBN-13: 978-1-60107-692-2
Library of Congress Control Number:
2011929106

Illustration and photo credits given in the captions. Front and back cover photo by Steve Quirt; design by Robin Walton.

To simplify information, trade names of products have been used. No endorsement of named or illustrated products is intended, nor is criticism implied of similar products that are not mentioned or illustrated.

UC PEER REVIEWED This publication has been anonymously peer reviewed for technical accuracy by University of California scientists and other qualified professionals. This review process was managed by the ANR Associate Editor for Farm Management and Economics.

Printed in the United States on recycled paper.

2500-pr-7/11-LR/RW

CONTENTS ✑

Preface. iv

Acknowledgments. v

1. Introducing the Farmstead and Artisan Cheese Business. 1
     Barbara Reed

2. Evaluating Your Resources: Is Cheesemaking for You?. 5
     Barbara Reed

3. Building a Business Plan 13
     Barbara Reed, Leslie J. Butler, and Ellie Rilla

4. Plant Layout and Design. 31
     Barbara Reed

5. Designing Your Marketing Strategy. 45
     Leslie J. Butler, Barbara Reed, Ellie Rilla, and Holly George

6. Risk Management 79
     Barbara Reed

7. Regulations 121
     Barbara Reed, Ellie Rilla, and Holly George

Glossary. 133

Measurement Conversion Table 141

Index 143

Photo: Todd Tankersly.

PREFACE

Just to be clear, this publication is not about cheesemaking, nor is it a statistical summary of the artisan and farmstead cheese industry. This publication is about the process of building an excellent cheesemaking business. It addresses certain features of cheesemaking procedures and food safety as a key part of good business management. While applicable to any area of the United States, the book also covers aspects of the cheesemaking business unique to the West, including marketing and distribution challenges and state regulatory issues.

This publication is written for readers who have no previous experience in food processing, dairy production, or business management. For readers with experience, some information contained in this publication will be review and will serve as an excellent refresher and desk reference. Although some of it can be found in similar forms elsewhere, the information here has been researched and uniquely assembled into a single document that is tailored for this market and regulatory climate.

The cheesemaking process itself has been covered very effectively in other publications such as *Cheesemaking Practice* by R. Scott, R. Robinson, and R. Wilbey (Kluwer Academic/ Plenum Publishers, 1998), *American Farmstead Cheese: The Complete Guide to Making and Selling Artisan Cheeses* by P. Kindstedt (Chelsea Green Publishing Company, 2005), and *The Farmstead Creamery Advisor: The Complete Guide to Building and Running a Small, Farm-Based Cheese Business* by G. Caldwell (Chelsea Green Publishing Company, 2010). In addition, the University of Guelph in Canada maintains a Dairy Science and Technology Web site at http://www.foodsci. uoguelph.ca/dairyedu/.

Navigating the start-up of any business is hard work, but cheesemaking has its own special challenges. Should you decide to run your own cheese business, most days you will be up to your elbows (literally) in curds and whey, and you will learn that there is never an end to washing equipment, dishes, vats, cheese presses, floors, walls, drains, and so on. On the bright side, there is nothing like taking fresh, sweet milk and turning it into a lovely wheel of nutty, aged Gouda or a wonderful, tangy Chèvre with herbs.

In my role as a University of California Dairy Advisor, I decided to write this publication because over the past several years I have spoken with many would-be cheesemakers who wanted to get started in the business. They made some nice cheeses at home or attended hands-on cheesemaking classes, but they had a hard time finding out about the business side of cheesemaking. This clientele was bright and very motivated but had no idea of the complexity of the business beyond the cheese vat.

I hope this publication informs and enlightens you, so that you can avoid some of the pitfalls of others who have gone before you and don't have to reinvent the wheel! With luck and hard work, you can create the best cheesemaking business ever. Or, you may see that the complexity of this business is not a good match for your skills, interests, and risk tolerance—and you move on to something else. It doesn't mean you have to stop loving cheese!

I want to thank all the farmstead and artisan cheesemakers in California who have shared with me about the challenges, problems, and successes they have had in their work. California is home to some of the best cheesemakers in the world, and it is a privilege to work with them.

Barbara Reed
Master of Architecture, University of Oregon
LEED AP

ACKNOWLEDGMENTS

Farmstead and Artisan Cheeses: A Guide to Building a Business was possible because of the generous contributions from many people in California whose lives are dedicated to great cheese.

For their support and encouragement, the authors thank all the California farmstead and artisan cheesemakers who participated in the University of California Cooperative Extension workshops, short courses, surveys, and interviews that became the foundation material for this book.

For their assistance in the development and implementation of the specialty cheese marketing and consumer research projects, the Farmstead Cheesemaking Workshops, and the Food Safety Training for Cheesemakers: Prerequisites for HACCP Workshops, the authors thank Dr. John Bruhn, Food Science and Technology Emeritus Cooperative Extension Specialist, Dr. Christine Bruhn, Consumer Food Marketing Specialist, Dr. Linda Harris, Cooperative Microbial Food Safety Extension Specialist, and Dr. Shermain Hardesty, Agricultural and Resource Economics Extension Specialist at the University of California, Davis, Dr. Nana Y. Farkye, Dairy Science Faculty at the Dairy Product Technology Center at Cal Poly, San Luis Obispo, Professor Cindy Daley at the College of Agriculture at Chico State University, and Mary Rumiano of Rumiano Cheese.

In addition, the authors thank several state organizations and educational institutions for their support and for providing permission to adapt their resources. Credit has been given to original sources at the end of specific chapters and in the "References." Thanks also to Jennifer Bice, Redwood Hill Farm and Creamery, along with Sue Conley and Peggy Smith, Cowgirl Creamery, for their review and edits of various chapters.

For sharing energy modeling and architectural design information, the authors thank Petra Cooper and Lapointe Architects.

Special acknowledgment goes to Holly George, Livestock and Natural Resource Advisor, Plumas-Sierra Counties, for assistance with the chapters on building a business plan and regulations.

For photographs, the authors thank Mike Poe, University of California, Davis, and Steve Quirt, UC Cooperative Extension, Marin County. Additional thanks for sharing photographs go to the Agricultural Institute of Marin, Bellwether Farms, Sharon Bice, Cowgirl Creamery, Point Reyes Farmstead Cheese, Redwood Hill Farm and Creamery, Curtis Meyers, Michael Suarez, Todd Tankersly, Valley Ford Cheese Company, and Carleen Weirauch.

Photo credit: Cowgirl Creamery

Farmstead and Artisan Cheeses

A GUIDE TO BUILDING A BUSINESS

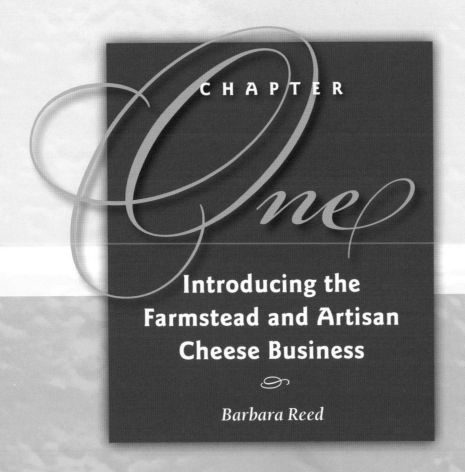

CHAPTER

One

Introducing the Farmstead and Artisan Cheese Business

Barbara Reed

Artisan and farmstead cheeses are a part of U.S. cheese
history, and here in the West they go all the way back to
the priests of the California missions. But unlike in Europe,
the Middle East, and other regions that have thousands
of years of cheesemaking tradition, U.S. cheese
production evolved in just the past 150 years.

In California it began with the missions, and then the newly settled homemakers helping to feed the miners, and continued with the cheesemaking traditions of Dutch and Italian immigrants and more Americans moving west. When the gold rush came along in the mid-1800s, the food and cheese demands of the miners pushed production upwards. Other places in the United States have similar stories. Large-scale cheese production in the United States eclipsed farmstead and artisan production before a great diversity of local cheeses was able to develop to the same extent as in Europe or in nations with a longer dairy foods history. With the advent of World War II and the need to further gear up the country's food supply, small-scale production of unique and diverse cheese virtually disappeared, especially in the West.

From the 1950s to the 1990s, commodity production dominated the U.S. cheese market, with a few exceptions. In Wisconsin and California, in areas of New York, and in Vermont, cheese production grew by leaps and bounds and became the dominant product made from milk. More than a third of the world's cheese is made in 450 plants in the United States. Today, "mainstream" cheese production in the various parts of the United States occurs on a grand scale, with gleaming industrial plants

pressing billions of pounds of cheese out of 50-foot towers, making products that go onto everything from pizza and tacos to cheese puffs.

In the 1990s, cheesemakers began an evolution similar to that of the wine industry in the 1970s. It took almost 30 years for California to evolve from the jug wine capital of the United States to a state known for its fine wines. The wine industry built its reputation by creating products on a small scale that were equal or superior to European wines in the same price categories. Although the artisan cheese industry is still in its adolescence today, like the wine industry of the 1970s, it has lots of room for growth. In 2009, the American Cheese Society competition had over 1,300 entries from around the nation.

This growth in the specialty cheese business is attracting new farmers and specialty food entrepreneurs to the market. The American Cheese Society membership reflects this trend. The Society is an educational resource for North American cheesemakers and the public. Members share knowledge and experience in cheesemaking as a hobby or as a commercial enterprise, with special attention given to specialty and farmhouse cheeses made from all types of milk, including cow's, goat's, and sheep's milk. Membership has grown from just over 400 members in 2001 to 1,200 members in 2009, with most of the growth occurring since 2004.

Many specialty cheeses are described as artisan cheeses. This means they are made by hand in limited quantities, using traditional methods with great attention to detail, and possessing unique taste and quality characteristics that differentiate them from other cheeses. For some specialty cheeses such as Feta and goat cheese, U.S. sales volumes have grown 20 percent annually. When cheese is made on the farm and with milk from the farm, it is referred to as farmstead cheese. The California Milk Advisory Board (CMAB) surveyed more than 50 leading restaurants in San Francisco and the Napa and Sonoma Valley wine region about artisan cheese use in 1995 and again in 2000. Back in 1995, none of the restaurants surveyed had a cheese course and few artisan cheeses appeared on the menu. By 2000, nearly half of the restaurants had artisan cheese on the menu as an ingredient and two-thirds of the restaurants featured cheese courses. Artisan cheeses are also making inroads into specialty retailers such as Whole Foods and Dean and Deluca.

Vertical integration of dairy production and specialty cheese processing allows dairy farmers to set prices that actually cover their costs of production, rather than selling raw milk at a price they don't control. Some dairy producers have capitalized on the expansion of this market and are creating unique, handmade cheeses in small-scale operations following European traditions. These farmstead cheeses are made with milk from the farm, on the farm. According to the CMAB, although the current U.S. consumption of cheese is estimated at just over 30 pounds per person (a 20 percent increase from 1990), consumption in this country still lags behind that in traditional bastions of cheese consumption like France, where the average annual figure is 54 pounds per person (fig. 1.1). It is expected that total cheese consumption in the United States will grow to about 40 pounds per capita by 2016. Even though specialty cheese constitutes only about 10 percent of all the cheese consumed nationally, specialty cheese consumption on a per capita basis has grown five times faster than total cheese consumption over the past decade. Some California specialty cheesemakers interviewed for a marketing study have grown their businesses by more than 40 percent annually in each of the past few years. As a category, specialty cheeses already make up about 13 percent of California's 1.8 billion-pound cheese industry, which is about double the national average. For additional consumer data on specialty cheese consumption in the United States, see table 1.1.

The bottom line for potential cheesemakers is that the market for specialty cheeses is continuing to grow, but so is the competition within the category. You need to thoroughly research your market and understand the marketplace before you get started.

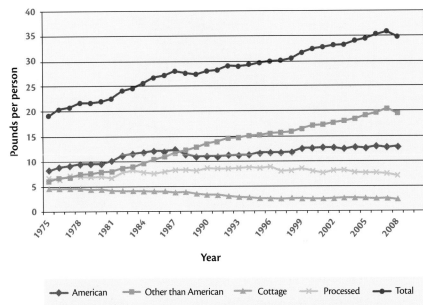

Figure 1.1. U.S. per capita cheese consumption between 1975 and 2008. While total cheese use has trended upward, most of that growth is due to increasing consumption of other-than-American cheese. The USDA expects growth to level out as population demographics change.

U.S. Per Capita Cheese Consumption 1975–2008

American cheeses are Cheddar, Colby, Monterey, and Jack. Other-than-American cheeses include Mozzarella, Parmesan, Provolone, Ricotta, Swiss, Hispanic varieties, and others.

Source: Adapted from *Long-Term Growth in U.S. Cheese Consumption May Slow,* USDA ERS report, LDP-M-193-01.

Table 1.1. Growth in specialty cheese* consumption in the United States, 1994–2003

Type of cheese consumption	1994	2003	Growth
Total cheese consumption (million pounds)	7,000	8,800	1,800 (+26%)
Specialty cheese consumption (million pounds)	420 (6% of total)	815 (9% of total)	395 (+94%)
Total cheese consumption per capita (pounds)	26.6	30.6	4 (+15%)
Specialty cheese consumption per capita (pounds)	1.6	2.8	1.20 (+ 75%)

Source: California Milk Advisory Board 2004.

Note: *For purposes of the study, *specialty cheese* was defined as "natural cheese that commands a higher price than a commodity cheese because of its high quality, limited production, and value-added production techniques or ingredients." This includes varieties commonly designated as specialty cheeses, commodity-type cheeses aged 12 months or longer (i.e., Cheddar, Jack), cheeses flavored with vegetables, fruits, or herbs/spices, and Mozzarella packed in water or oil, less than 2 weeks old.

Parts of Chapter One adapted from Reed, B. and C. M. Bruhn. 2003. Sampling and Farm Stories Prompt Consumers to Buy Specialty Cheeses. California Agriculture 57(3): 76–80.

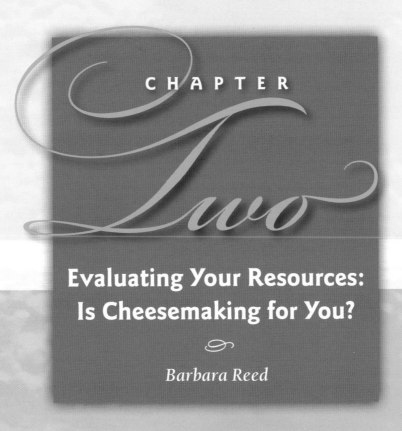

CHAPTER Two

Evaluating Your Resources: Is Cheesemaking for You?

Barbara Reed

Chapter Goals 6

A Good Plan Is a Powerful Tool. 6

Identify Your Competition and Your Customers 7

Assess Your Skills. 7

Financial Resources 11

Physical Resources. 11

Points to Remember. 11

CHAPTER GOALS

The goals of this chapter are to help you

- determine if a cheesemaking enterprise fits your interests, philosophy, and goals

- consider the costs and benefits of a cheesemaking enterprise, especially if vertically integrating your dairy farming business

A GOOD PLAN IS A POWERFUL TOOL

Meticulous planning, energy, patience, technical skills, and a fundamental understanding of food safety are needed to start and run a cheese business. So, before you buy your first cheese vat, take time to plan well. Planning helps you correct mistakes while they are still just words or pictures on paper, and there is no risk if you need to go back to the drawing board a few times.

First, you will need to determine whether a cheesemaking enterprise fits your interests, philosophy, and goals. If you are currently in dairy farming, what are your operation's current resources and outlook? Will a cheesemaking business add to that, or will it make it more challenging?

You need to discuss your business and personal goals with your business partners and/or family. Everyone needs to be in agreement and take part in the decision making before you move ahead.

Artisan and farmstead cheesemaking is *not* commodity production. When you vertically integrate your business, you will become actively engaged in the development, manufacturing, marketing, and sales of your product. It will no longer be enough to milk the cows (or goats or sheep) and tend to the farming operations. You will be creating an entirely new business that should be run as its own enterprise. If your farm is financially marginal now, you will have to ask yourself

"Try to work in an artisan cheese business to see what it's really like. Cheesemaking is labor intensive, hard work physically, and it must be attended to on weekends and holidays, much like animal care."

Cheesemaker Story

some tough questions about the financial and personal demands of adding a new enterprise.

As you assess your interests, philosophy, and goals, here are some questions to ask:

- Will starting a new business give our family adequate personal and family time? Are we good at setting limits so that we can still be a family and manage new business demands?

- Do we have strong enough family bonds and communication skills to hold the family together even if we experience future disagreements about the business?

- Do we have the labor and skill sets within our family to create a new business, or will we need to hire additional help?

- Which family members are willing and able to be a part of the new business?

- Is this new business going to be passed on to heirs or sold?

- What are our biggest weaknesses and resource gaps, and how can we remedy those issues?

- Are we "buying ourselves a job," or should we make some other kind of investment with our resources?

- Will we create a business that can support us, or will we be working "for free"?

- Have we set specific revenue goals for the business?

- What is our core mission and philosophy?

- What level of financial risk can we tolerate?

- Do we have assets we can commit to raise necessary business capital?

- Do we have a plan for growth that matches our resources and abilities?

- Do we have a succession plan if there is an unexpected loss of a family member or business partner?

While some of the questions asked in this chapter may seem very direct, it is important to answer them honestly. There should be a good fit between the business and your interests, philosophy, and goals. A little bit of unvarnished truth now can help you be very successful later. For some examples of the benefits and costs of a cheese business, see table 2.1.

IDENTIFY YOUR COMPETITION AND YOUR CUSTOMERS

Learn about your product niche and its customers. Go to chain grocery stores, specialty grocers, cheese shops, farmers' markets. Go online and look in specialty food catalogs. Look at the cheese types, the typical sale weight (if prepackaged), the type of packaging, and the retail price. Are there any sales promotions on the items? Are the cheeses imported or domestic? How many cheeses are already in the category that might fit your new cheese?

Attend the Fancy Food Show in San Francisco, held every year in January, or the summer show on the East Coast. It is one of the largest specialty food shows in the nation. Attend the American Cheese Society (ACS) annual meeting in August. It is the largest gathering of cheesemakers and one of the largest cheesemaking competitions in the country. Join the ACS and subscribe to their newsletter. Join a local cheesemakers' group or a state guild such as the California Artisan Cheese Guild. There are cheese councils or guilds in California, Oregon, Pennsylvania, Vermont, Washington, and Wisconsin. Visit all the local cheesemakers in your area who open their plants to visitors. Read trade publications like the *Cheese Reporter, Cheese Market News,* and *Culture Magazine.* For more suggestions about how to find your ideal customers, see figure 2.1.

ASSESS YOUR SKILLS

When you review the skill set of your family and others with whom you will be going into business, you may come to the conclusion that being the cheesemaker in the cheese business is not for you. Your time and energy may be more appropriately spent in product development, marketing, or strategic management of the business. It is not productive to perform a job you don't like because you don't have the right skill set or temperament. Figure 2.2 offers a short skill-assessment exercise for you and those who will work with you.

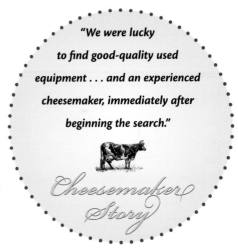

"We were lucky to find good-quality used equipment . . . and an experienced cheesemaker, immediately after beginning the search."

Cheesemaker Story

Table 2.1. Benefits and costs of a farmstead or artisan cheesemaking business

Benefits	Costs
Provides potential additional income.	Requires substantial investment in infrastructure and equipment.
Creates a business that will grow in value.	Adds work and responsibilities to the family.
Allows you to be your own boss.	Becomes a highly regulated business with many agencies involved.
Is personally rewarding.	Exposes you to additional risk and liability due to food manufacturing, and you can be held personally liable.
Allows you to connect more closely to your customers.	
Allows you to create a signature product.	Demands your full and constant attention, interfering with family time and activities.
Provides you with an opportunity to tell your story and educate your customers about family farms.	May require you to delete or change products to fulfill consumer preferences, since the customer is always right.
Can help support local agricultural policies.	

Figure 2.1

Who Are My Ideal Customers?

Who is in my customer base? _____

_____

_____

Are my customers local, regional, or national (or a combination)? _____

_____

_____

What are their ages and income?_____

_____

_____

Do they consider themselves food lovers and are they willing to try new foods? _____

_____

_____

_____

_____

How much disposable income do they have for discretionary purchases like wine and cheese? _____

_____

_____

Where do they shop and how much do they spend on groceries? _____

_____

_____

- ☐ Are they concerned about family farms and locally produced foods?

- ☐ Do they buy organic foods?

- ☐ Are they concerned about the potential health benefits of certain foods?

- ☐ Are they interested in the story of the food's production and the people who produce it?

Photo: Michael Suarez.

Figure 2.2

Consider Your Operational and Management Skills

What personal attributes do I bring to a cheesemaking operation? Am I good at business and financial management, marketing and sales, food technology, quality assurance, and logistics?

Have everyone who will participate in the business fill out a skills-assessment form such as the one below.

Skill-assessment question	No	Somewhat	Yes
Do I like doing work that is physically hard?			
Do I like working indoors in situations requiring sanitary conditions, special clothing, hair nets, no jewelry, etc.			
Do I like working in a manufacturing environment full of water, milk, cleaning solutions, and food products?			
Am I • detail oriented? • organized? • persistent? • able to make and carry out decisions? • willing to accept responsibility for other people's health and safety?			
Am I willing to work long hours in sometimes difficult situations?			
Do I have in-depth knowledge of cheese technology?			
Do I have training in food manufacturing, microbiology, or quality assurance?			
Do I have experience with • budgeting and cash flow? • managing people? • effective communication? • keeping detailed (manufacturing) records?			

For the categories where you and other partners may answer "no," consider whether you can hire personnel with these strengths. What is an alternate strategy to solve the "no"?

"My advice to someone getting started is to hire consultants to help you where you don't have the skills."

Cheesemaker Story

Figure 2.3

What Are My Physical Resources?

Land

How much land do I have available for a cheese plant on my current farming site? How might this use conflict with current uses, and how can potential conflicts or disruptions be minimized? If I'm planning on-site retail, how will this fit with my current activities? _____

_____

_____

Terroir (a sense of place)

What elements of my farm help create a unique environment in which I can produce a special food product? What breed(s) and specie(s) of dairy animals do I have? How do the topography, soil type, elevation, and climate contribute to the local plant community? How do my animals use this plant community? _____

_____

_____

Infrastructure

What infrastructure do I have in place? Is there adequate electrical, gas (propane or natural gas line), and water supply (municipal or well)? Do I have sewage or septic hookup, and what is the capacity of the disposal system? _____

_____

_____

Character of the farm

If I am considering farm visits or a retail outlet at the farm, what is the condition of the other farm buildings and grounds? What would I have to do to make the facility attractive and safe for regular visitors?

_____

_____

Proximity to transportation routes

How far is my property from a major interstate highway? What is the condition of access roads between my farm and the highway? If I am considering a farm store, where does my client base come from? Am I on a well-traveled local road or off the beaten path? Are there any specialty food distributors that serve my immediate area? If not, how close do they come to my location? If I am renting or purchasing land, how does it fit the above considerations? _____

_____

_____

Local attractions

Are there other food artisans or regional food-marketing groups in my area that help attract clientele to the region?

_____

_____

Support services

Can I get commercial garbage service at my location? Are there any professionals in my area who specialize in equipment for dairy food processing? _____

_____

_____

Regulations

Have I checked out all the necessary regulations? Can I meet the letter of the law?

_____

(See Chapter Seven for more information on regulations.)

FINANCIAL RESOURCES

You will need money for business start-up and operational costs until your business has a positive cash flow. Building and equipping a cheese plant and getting your product to market can require substantial capital. Do you have cash savings? If not, do you have access to capital through traditional banks, family, or other private lenders? What collateral do you have to secure those loans? In the next chapter we will talk more about borrowing money to start a business and what bankers look for when they consider making a loan. Just remember, even the best business plan won't impress a bank officer if you expect to borrow 100 percent of your start-up costs and you have no collateral of your own to commit to the project.

PHYSICAL RESOURCES

Unless you are buying milk and leasing a facility, you will need to consider how your physical resources will contribute to your efforts to start a new business. Figure 2.3 provides you with a list of questions to review about your local and regional resources.

POINTS TO REMEMBER

→ Complete and detailed planning is critical to the success of a new cheesemaking enterprise.

→ The enterprise should be a good match with your goals, philosophy, and character, as well as those of your family or business partners. And if you are farming now, the new enterprise should be compatible with the existing business.

→ Know and understand your competition, the market, where your product will fit, and how it will be competitive.

→ Cheese production is repetitive work that requires attention to detail, technical knowledge, and great concern for food safety and the welfare of your customer.

→ You need to be ready to commit personal resources, both spiritual and financial, to make the project work.

Sections on evaluating your resources adapted in part from University of California Cooperative Extension Farmstead Cheesemaking Workshops, 2003–2005. Glenn County Cooperative Extension, Orland, CA.

Terroir in the United States is defined as "a sense of place," referring to subtle local influences in the flavor of foods and wine. If there is something special about your soils, be sure to share it. *Photo:* Steve Quirt.

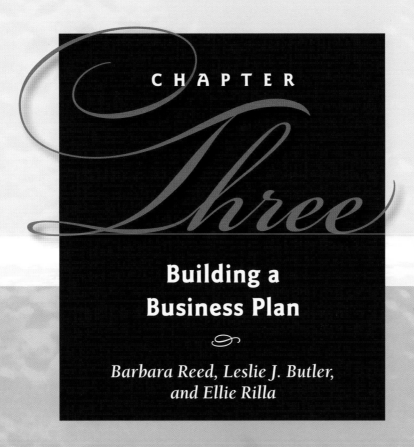

CHAPTER
Three

Building a
Business Plan

*Barbara Reed, Leslie J. Butler,
and Ellie Rilla*

Chapter Goals 14

Financing Your Business 14

Business Planning 15

Set Measurable Goals and Objectives 20

Low-Cost Start-Up 24

Business Succession 25

Points to Remember 25

Business Plan Resources 25

References 27

Appendix 3.A. Business Plan Outline 28

FINANCING YOUR BUSINESS

Before you daydream about accepting the award for "Best of Show" cheese at the American Cheese Society annual meeting, you will have to invest a substantial amount of money to realize your dreams. There are two main reasons for this. First, dairy regulations are stringent, so inspectors are not going to approve a food-manufacturing facility in your home kitchen or garage. Second, you will need to design a facility that allows you to make and sell a large enough volume of cheese to create a cash flow for the business and to pay yourself a salary for the first year. In this publication, we assume a business will be created to produce income and that it will have a positive cash flow. We do not take into account off-farm income or "hobby cheesemaking" enterprises. We will discuss microenterprises in Chapter Five, "Designing Your Marketing Strategy."

Conservatively, these costs will range from $500,000 to $750,000, not including land.

This covers construction of a modest facility with sufficient space for making cheese, aging, packaging and storing inventory, equipment washing, chemical and equipment storage, and employee changing facilities and bathrooms. Facility design should promote good manufacturing practices that will help control personnel traffic and prevent contaminants from getting into your facility. It is possible to lease land, possibly a former dairy, and convert a barn or agricultural buildings to a cheese plant, albeit small, for less cost. Some creative start-ups have used portable buildings on trailers or shipping containers to build their milking parlor, as well as make and aging rooms, to keep costs down to a range of $75,000 to $150,000 (Rilla 2011).

You will need to commit some personal financial assets to create your business. All the good business planning in the world won't get you anywhere if you don't have any assets or collateral to commit to the development of your business.

Banks use a matrix to determine the risk they face in lending to you. With the recent events in the economy, these rules are tighter than ever. Here are some of the criteria they may use to evaluate their risk of lending to you:

- Will you pledge available collateral?

- Do you have previous experience in this type of business?

- Is the business currently in operation, or is it a start-up?

- If you have agreed to pledge collateral, can you contribute 10 to 20 percent of the total

"Be well capitalized and be prepared to lose money for a while. It takes longer and costs more than you think."

Cheesemaker Story

equity needed for the project? The lender may ask for a minimum 30 percent investment from you if the business is a start-up.

- Do you have a dollar of collateral for every dollar you plan to borrow? (If not, this will constrain your borrowing.)

- Do you have orders from customers (in writing) that you can show to the bank officer?

Small business lenders give loans for fixed assets (like buildings or land), equipment or machinery, and lines of credit for accounts receivable and/or inventory. Lenders will immediately discount the value of new equipment by 50 percent when considering its value as collateral. Discuss with your bank the matrix used to evaluate your credit worthiness. Find out what you may be able to borrow before your plans outgrow your financial resources. You may also borrow from friends and family rather than a commercial bank. They may have better interest rates and repayment terms. Regardless of your lending source, get agreements in writing, make sure you understand the agreements you sign, and have the documents reviewed by a lawyer.

There are a number of other financing sources besides bank or personal financing. Debt financing, equity financing, traditional agricultural lenders, small farm–friendly banks and holding companies, Small Business Administration (SBA) and rural economic development agencies, councils, and districts are other sources of potential financing.

Using any of these sources entails risk. Borrowing inherently brings risk, either financial or personal. The various types of financing are described in the following pages.

Debt Financing and Equity Financing

Debt financing is financing in which you borrow money or take on debt to further your business. You still own your business, and you still make all of your business decisions. Equity financing means you sell a piece of your business. You no longer own the entire business, and you usually are accountable to other people when making decisions. Equity financing includes limited partnerships or stock offerings, both of which require professional legal advice and assistance.

Traditional Agriculture Lenders

Traditional farm lenders include the Farm Credit Service and the Farm Service Agency. The Farm Credit Service is a collection of federally chartered, borrower-owned credit cooperatives (the "Farm Credit Cooperative Banks"). They lend to agricultural operations and provide rural home loans.

Another traditional agriculture lender is the Farm Service Agency. This agency is part of the U.S. Department of Agriculture. It has a direct-lending program and a loan-guarantee program, providing money when other lenders won't. For a comprehensive listing of federal resources and programs, see the USDA publication *Building Sustainable Farms, Ranches and Communities* (Krome, Maurer, and Weid 2009).

Small Business Administration

The Small Business Association (SBA) offers two primary loan programs that provide funding to small businesses unable to obtain capital through normal lending channels. These programs are the 7(a) and 504 loan-guarantee programs. The 7(a) and 504 programs are targeted mostly to real estate and fixed-asset financing, although the 7(a) program can also make loans for working capital.

SBA loans are made through private lenders. Many rural and small-town banks are able to make SBA-guaranteed loans. The SBA itself has no funds for direct lending or grants.

Rural Economic Development Agencies, Councils, and Districts

Rural economic development agencies oversee, distribute, and lend monies from federal and state community development block grants and from USDA rural development agency lending programs. Their funds are often tagged for specific purposes such as job creation or retention, housing rehabilitation, rural infrastructure tied to increased employment, micro loans to start-up businesses with job creation potential, and rehabilitation of community facilities.

BUSINESS PLANNING

Don't attempt to start and run a business without a plan. Most banks will require a business plan as part of a loan application package. In California, a third of North Bay cheesemakers interviewed in 2010 had developed business plans for their cheese operation and had sought funding for cheese plant development or expansion (Rilla 2011). Even if you don't need financing, you should prepare

"I could have used much more capital, [done] more work with a business plan, and [had more] vision for the long-term priorities."

Cheesemaker Story

a business plan. Think of it as a road map. You can choose many routes and change the course of your trip as you go along—but without a map, you don't know where you are, where you are going, or how to choose your next destination.

A business plan is a dynamic tool that will help you through the start-up process and will help you make well-informed business decisions. Writing a plan forces you to quantify your mission, your long- and short-term goals, and your values and business strategy. It also provides you with an organized system for testing your ideas. The document becomes a guide against which you can measure your progress,

reflect on your success, and make adjustments when things don't go as planned. Studies have shown that businesses that take the time to do a business plan increase their chances of success by around 70 percent.

Entire books are dedicated to writing business plans, and they are widely available. If you do an Internet search of "business plan books," you will get over 150,000 hits. You can purchase a computerized business plan template for $100 to $200, hire someone to assist you in the preparation of your business plan, or buy a how-to book for under $50.

In this chapter, we will address a few of the most critical aspects of the business plan. But because this publication is an overview of the entire development of a business, we won't attempt to create a template for you here. Small Business Development Centers in your community can help, or you can find online templates for free. Small Business Development Centers (SBDCs) are located in all 50 states and are partnerships primarily between the government and colleges/universities. They are administered by the Small Business Administration. SBDC professionals can provide assistance to you as a small business owner and aspiring entrepreneur.

You can also find a list of business plan resources at the end of this chapter and a typical business plan outline in appendix 3.A. Regardless of

how you complete your plan, however, no one can develop your strategies and goals for you. You will have to create the overarching principles upon which you build your business.

Executive Summary

The executive summary is a one-page summary of your plans. The first part consists of your business idea, and the second consists of the conclusions you've provided in your financial strategy. The executive summary comes first on your business plan, but it is written last.

Mission Statement

A mission statement merges your personal philosophy with your business goals. It should explain your purpose, your business, and your values in a brief statement. Hopefully you spent some time completing the exercises from Chapter Two, "Evaluating Your Resources: Is Cheesemaking for You?" You, your family, and your partners have already written down the values important to you,

Be sure to share the story of your farm and animals with your customers. *Photo: Steve Quirt.*

discussed your commitment to a new enterprise, revealed some strengths and weaknesses, and can see where potential conflicts might arise.

Because you will use the mission statement as a guiding principle in your business, the next step will be for you, your family, and your partners to distill those previous discussions of your values and goals into a concise and accurate reflection of what you hope to accomplish. This will be especially important if the business will place new demands on family members and bring customers to your farm. Refer to figure 3.1 for a sample mission statement. Use figure 3.2 for establishing and aligning your priorities. Use figure 3.3 as a template for drafting your own mission statement.

Your product:

- What is it? Why should people buy your cheese? How is it unique? Why is it worth the cost? How does your cheese compare to others in its category? Is it so unique that it stands on its own?

Your target market:

- Who are your ideal customers? Define their gender, age, education, income, food preferences, hobbies, and interests. Where do these people shop? Do they value fresh food? Handmade food? Are they concerned about antibiotics and hormones in milk and

---

Figure 3.1

ABC Dairy Farm Sample Mission Statement

ABC DAIRY FARM

Background Information

This operation is a family-owned dairy, run by two generations of family members who live on the dairy. The dairy is grass-based and runs Jersey cattle in a year-round operation. The first generation of family members wants to retire from day-to-day operations, and two grandchildren (finishing college with business and marketing degrees) want to come back to the farm. The farm is located on a well-traveled road, near a well-known area for international and regional tourism. There are other direct-sales farms in the area.

The family is concerned that the commodity milk market does not value the care and quality put into the milk production on the farm. The family also wants to capture added value for its milk. The third generation believes that vertical integration of the operation into cheesemaking will ensure the farm's economic viability and its ability to capture some market share of the food-tourism dollars.

Selected new enterprises

- Value-added farmstead cheesemaking
- Direct-market sales of cheeses
- Membership on a farm trails listing and creation of seasonal cheese shop

Mission Statement

ABC Dairy Farm will preserve the heritage of our grass-based dairy farm by producing the highest-quality, handmade Cheddar cheese and selling it at our own farm store, farmers' markets, and specialty grocers. We will utilize the resources and talents of our family to educate people about the importance of locally produced food and the unique attributes of our farm, animals, and cheese production.

---

what is fed to the cow or goat? Are they focused on the quality and not the cost of food?

- How does your enterprise fit with others in the region?

- Are you the only local food producer in the area? Are there other farm businesses that have also vertically integrated, such as olive oil producers, grass-fed beef ranchers, and vegetable growers with farm stands? Will you have competitors or collaborators?

The above aspects of your plan are covered in depth in Chapter Five, "Designing Your Marketing Strategy."

Figure 3.2

Our Priorities

Values	Least important	Somewhat important	More important	Most important
spending time with our family				
enjoying our peace and privacy				
keeping the land in our family				
producing locally grown/made food				
providing the highest-quality product				
providing a unique product				
teaching others				
preserving our story and identity in the food we produce				
continuing a family business				
ensuring our economic sustainability				
ensuring a secure retirement				
Use blank space to insert additional values you might have.				

Business Concept

You may have heard of speed dating and speed interviews. In addition to writing about your products and services in your business plan, you should discuss your overall concept. Write it as though you had only 2 minutes to summarize it for a banker, an advertising firm, or someone in an elevator. Your ideas have to be persuasive, simple, and complete. Your overall business concept must be as clear as your mission statement.

If you decide to sell to the public at your creamery, design the facility so that your customers can see cheese being made. *Photo:* **Steve Quirt.**

Figure 3.3

Our Mission Statement

Background Information

_____
_____
_____
_____
_____
_____
_____
_____
_____
_____

New Enterprise(s)

_____
_____
_____
_____
_____
_____
_____
_____
_____
_____

Mission Statement

_____
_____
_____
_____
_____
_____
_____
_____
_____

Example of a business concept

A sample description of Sophie's Chèvre business concept is provided in figure 3.4. Not only does Sophie's raise purebred goats, but the business also makes cheese, sells breeding stock, and conducts educational workshops. Use figure 3.5 for drafting your own business concept.

SET MEASURABLE GOALS AND OBJECTIVES

As mentioned in Chapter Two, the choices you make about your business operation will shape every aspect of your planning process and assumptions. You have to narrow your choices and set measurable goals and objectives in order to create a realistic list of tasks and actions.

Figure 3.4

Sophie's Chèvre Business Concept

Sophie's Chèvre will be a vertically integrated goat dairy in Laurel Creek, about 15 miles from Ookinsville. We will operate the foothill farm sustainably, utilizing the lower irrigated hill area for grazing and the upper oak woodland hills for browsing. We will operate the farm as an S Corporation with help from students who participate in an international agricultural apprentice program.

We will raise registered alpine dairy goats that will be classified through linear appraisal and production tested through the American Dairy Goats Association and the California Dairy Herd Improvement Association.

We will operate the dairy and cheesemaking on a seasonal basis, milking the goats and making cheese between the months of February and October. We will generate revenue from cheese production; the sale of registered does and doe kids for dairy production; the sale of buck kids into the ethnic meat market; and educational workshops on cheesemaking. We will advertise cheese classes on our Internet site and host 4-day educational sessions twice a year.

Our cheese business will be 80 percent Chèvre and other non-aged, traditional goat cheese (Feta, Crottin, etc.) and 20 percent aged cheeses (Gouda style). Our market priorities are Northern California specialty grocers, farmers' markets, and high-end foodservice.

We will finance our start-up by borrowing against the equity in our ranch and by using retirement income from previous federal employment.

Figure 3.5

Our Business Concept

Enterprise Name: _____

Business Concept

_____

_____

_____

_____

_____

_____

_____

_____

_____

_____

_____

_____

For each objective, you will have to make a decision that will affect your business concept. The following are sample goals and objectives for the ABC Dairy, starting with their mission statement. (Table 3.1 gives some detailed examples of developing tasks from a list of objectives for ABC Dairy, a fictional business.)

Sample Goals and Objectives

Mission statement

ABC Dairy Farm will preserve its heritage as a grass-based dairy farm by producing the highest-quality, handmade Cheddar cheese and selling it at our own farm store, farmers' markets, and specialty grocers. We will use the resources and talents of our family to educate people about the importance of locally produced food and the unique attributes of our farm, animals, and cheese production.

Sample goals

Goal 1 (accomplish in 2 years): Develop a market that appreciates the excellent quality of our product, the story of our family, and the unique attributes of our farm so that we achieve gross revenues of $500,000 from direct sales.

Expected outcomes

The goal will be satisfied when

- consumers seek out our product because of its quality and the unique attributes of our brand

- consumer demand for our cheese

drives the cheesemaking schedule to 7 days per week at the existing herd size of 50 cows

- gross sales match cheese output at 90 percent of existing milk production and cheese production capacity

- we have 100 percent penetration into local farmers' markets and specialty retail within a 90-mile radius, and on-farm sales reach 25 percent of total gross revenue

Objectives and action steps
(how, who, what, when)

- Develop a marketing narrative that is based on the story of our family, the unique attributes of our farm, and the quality of our product. Hire a professional designer to create a product logo and label, point-of sale materials, and other promotional literature (including Web site design). Budget this as a one-time expense.

- Commit 20 percent of operating budget to marketing in each of the first 2 years of operation to cover cost of "free" product, additional labor for start-up, and services of a cheese broker for specialty grocery store sales.

- Plan labor and production schedules to serve 50 percent of the region's farmers' markets in the first 6 months and 100 percent of the region's farmers' markets in the first year.

You may need to hire a milker if you become the cheesemaker or hire a cheesemaker so you continue with farming operations. If you try to do both, it may not go well. *Photo: Steve Quirt.*

Table 3.1. How to move from ideas to actions

Objective	Potential decision	Business concept	Action step	Task (partial list)
Combine some elements of agri-tourism into our cheese business.	Manufacturing only OR farm store and open to public?	Create farm store open to the public.	• Outline budget and labor for retail operation.	Determine store size, equipment needs, set-up costs, store hours, sales estimates, overhead, and operating costs. (Set up as separate enterprise from cheese manufac-turing to see if it creates cash flow.)
Capture added value from dairy enterprise.	Renegotiate milk contract, get organic certification, go into farmstead processing?	Continue existing dairy farm operations.	• Set up separate enterprises for cheesemaking and milk production. • Determine fair market price (opportunity cost) for milk purchased from farm.	Calculate current farm cost of production for milk, including a return to management. Build that milk price into operating budget and cash flow estimates for cheese operation.
Have cheesemaking expertise at farm.	Train family member(s)? Hire employees?	Plan to hire chief cheesemaker.	• Develop job description and salary estimate. • Start job search.	Write down key tasks for position, and desired skill level and experi-ence. Contact other cheesemakers to find out comparable pay rate.
Create a business that has some downtime.	Seasonal OR year-round production?	Produce cheese seasonally.	• Create cheese production schedule. • Anticipate effects on cash flow.	Create flow chart for daily produc-tion schedule; estimate number of person/hours for each process; and reconcile with milk supply and sales estimates.
Create a cheese that is aged less than 60 days.	Raw OR pasteurized milk?	Include pasteurizer in process to meet legal requirement for cheese aged less than 60 days.	• Build cost of pasteurizer into equipment budget. • Price new versus used. • Obtain pasteurizer license.	Get prices and specifications on equipment; study and take test for pasteurizer operator's license; and get pasteurizer certified by dairy inspector.
Create a product that will vary by season.	Whole un-standardized OR standardized milk?	Use whole milk with seasonal variation.	• Learn about effect of milk composition on cheese yield, attributes, and quality. • Examine seasonal changes in feed supply and understand effects on milk composition.	Take cheese technology class and/or hire qualified consultant to evaluate milk quality and compo-nents. Send milk samples out to dairy laboratory for evaluation, and do informal taste/sensory evaluation of milk supply. Con-duct research and development, making test cheeses by season.
Decrease calf feed costs.	Whey disposal: sell OR use?	Feed whey to calves. (Need to use sweet whey.)	• Estimate whey generation versus consumption. • Create back-up plan for whey utilization and disposal.	Build whey use into a feed budget for livestock; work out price and transport plan for whey feed sales; talk with dairy inspector and Regional Water Quality Control Board about whey effluent.
Use cultures in small quantities that are easy to store and handle.	Bulk cultures OR freeze-dried cultures?	Use freeze-dried cultures.	• Do research on culture variety and cost, as well as culture houses. • Determine how cultures match the type(s) of cheese we intend to make.	Take cheese technology class and/or hire qualified consultant for assistance with cheese develop-ment. Attend trade shows and contact sales representatives of culture houses. Test different cultures in the make process.
Capture as much revenue from sales as possible.	Direct sales OR wholesale?	Maximize direct sales at farm store, farmers' markets, and direct to retail.	• Develop estimates of –gross sales in each segment –labor and transportation costs –product pricing –additional permits, requirements, restrictions	Calculate net revenue for each cheese after deducting overhead and operational costs from aver-age sales price. Project optimum product and market mix.

- In the first 2 years of production, participate in every "food trail," Farm Bureau, and "locally grown or made" food event in a 50-mile radius. Agree to all requests for interviews and other forms of free press.

- Schedule regular tastings and other product promotions with specialty grocers for alternate months, and provide product information and training for their retail staff.

Goal 2 (accomplish in 5 years): Expand sales while retaining the excellent quality of our product, preserving the story of our family, and the unique attributes of our farm so that we achieve gross revenues of $1,000,000 from direct sales.

Expected outcomes
The goal will be satisfied when

- consumer demand for our cheese drives the need to expand the herd size to 100 cows

- gross sales match cheese output at 95 percent of existing milk production and cheese production capacity

- we have 100 percent penetration into local farmers' markets and specialty retail within a 90-mile radius, and on-farm sales reach 25 percent of total gross revenue

- we develop foodservice accounts when driven by demand

Objectives and action steps (how, who, what, when)

- Bring average direct-sale price to $17 per pound.

- Add three products in similar lines (two aged Cheddars and a wine-soaked Cheddar).

- Hire two more full-time employees.

- Design cheese room so production capacity can double in 5 years without substantial renovation and so renovation will not stop operation.

- Start with a single cheese-making vat, with plans to add another. (This way the cheesemaking process won't have to change but instead can just be duplicated. This is a simple alternative to changing vat size and having to "scale-up" the cheesemaking process to a larger volume.)

Finance Plan and Projections

Your financial strategy/statement is key to financing. It describes your existing debt and your financing needs, and it specifies your fixed assets, start-up costs, and cash flow forecasts as well as debt repayment schedules. Your financial statement also

explains how your new enterprise will fit into your current operation.

Financial documents to include in this section are your

- profit and loss statement with assumptions

- balance sheet including assets, liabilities, and net worth

- cash flow projections including sales projections and assumptions

As with creating business and marketing plans, the challenge of creating a budget and an assortment of financial statements can seem daunting. Yet, not only are they very important documents to create, they are especially necessary as a guide to your success in business. And in most instances they are required by bankers, tax advisors, and others involved with financing your business.

Help with your financial planning is available in many forms. You can look for a professional (usually locally) to do these tasks for you for a moderate fee. Check the Internet for firms who will do this. There are a number of simple and easy-to-understand computer programs that can generate all the documents you require. These programs give you the option of choosing which documents you need and the style in which you need them, while also providing the training you need to create them. An abundance of resources exists on the Internet that can help you understand these documents and generate them successfully. Many books are also available to guide you through this task. There is no question that these documents are necessary; the only question is which method you will choose to create them. See the end of this chapter for a sample listing of resources.

Supporting Documentation

You will furnish supporting documents at the end of the business plan. These include financial statements, customer support letters, sales contracts or potential sales agreements, a resume of your experience in the industry (if you have related experience), and credit terms available to your existing business.

In general, the work you do to plan your business will follow this sequence:

1. Product and market research

2. Feasibility decisions

3. Financing

4. Co-packing contract, subletting, or facility lease

   a. construction or remodeling of facilities

   b. equipment purchase and installation

5. On-site product research and development

6. Start-up

7. Product to market

Whatever timeline you establish for achieving the above steps, expect it to change. The hallmark of a good business owner is the ability to be flexible, anticipate change, and have an alternate plan ready if things don't go as expected.

LOW-COST START-UP

A Note on Research and Development and Breaking into the Marketplace

Most large grocery chains won't carry your product until you are an approved vendor. Before they will lend you money, most lenders want to see a letter from a customer (or potential customer) who has agreed to buy your product. You can't make enough cheese to secure the grocery chain business until you have borrowed some money and have a plant in which to make cheese. This is a dilemma, but it's one you can solve by approaching the challenges of market entry creatively.

Many of today's successful cheesemakers didn't build a facility the moment they got started. Some of them rented processing space at an existing cheese plant. Other cheesemakers paid another cheese company to start their product line (also known as co-packing), but provided raw materials, information about the desired cheesemaking process and product attributes, and spent time overseeing the cheesemaking and managing the marketing. They were willing to start small and sell directly

Selling to a supermarket usually requires that you become an approved vendor. *Photo: Barbara Reed.*

to consumers before venturing into large retail venues. Once their cheeses were established in the marketplace, selling well, and with brand recognition, the cheesemakers were positioned to go out on their own. At this point, they had the track record and cash flow to obtain start-up funding for their own plant. Then they could produce the volume of cheese needed to move into retail grocery venues.

BUSINESS SUCCESSION

In Stephen Covey's book, *The Seven Habits of Highly Effective People*, Habit #2 is "Begin with the end in mind." For Covey, this means planning for the life you want to have. In the cheesemaking business, it is about planning for the future of your business in the present.

A milestone in U.S. artisan cheesemaking was reached in 2006 when Laura Chenel of Sonoma, California, sold her goat cheese business to a French firm. Ms. Chenel was one of the first artisan cheesemakers in the United States. After 25 years in the business, she created an exit strategy and followed through with it. You may intend to pass your business on to other members of the family, sell to a partner, or sell to another company. Even if you can't predict how it will all work out

in the end, you should begin with the end in mind by planning for "what's next." This will help shape your business strategy and enable you to focus on how to reach your goals.

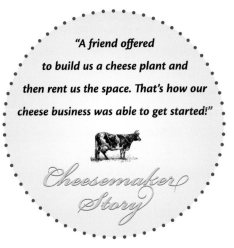

"A friend offered to build us a cheese plant and then rent us the space. That's how our cheese business was able to get started!"

Cheesemaker Story

POINTS TO REMEMBER

↪ A business plan is critical to your business's success. Remember, failure to plan is planning to fail.

↪ The business plan contains many components to be considered and written: the executive summary, mission statement, business concept or idea, measurable goals and objectives, background information (industry research and market analysis), management needs and management history, marketing strategy, financial strategy, and appendix.

↪ Creating a business plan allows you to anticipate your new enterprise's opportunities and challenges on paper— before you commit substantial resources.

↪ A business plan is essential if you want outside financing.

↪ To predict how your new enterprise will impact your entire operation, compare a financial statement of your new enterprise to a financial statement without it.

BUSINESS PLAN RESOURCES

Online Assistance

BizPlanIt. By the developers of BusinessPlanPro software, this site offers good, quick advice. BizPlanIt Web site, http://www.bizplanit.com/.

BPlans.com. This Web site offers a free sample marketing plan. BPlans.com Web site, http://www.bplans.com/.

Business Owner's Tool Kit. This Web site supplies information to help small business owners make decisions and "get the most from your business." Business Owner's Toolkit Web site, http://www.toolkit.cch.com.

Center for Business Planning. For helpful planning advice, see the Business Planning Web site, www. businessplans.org/.

MasterPlans. For information regarding consultants, see the MasterPlans Web site, www.MasterPlans.com.

My Own Business, Inc. This Web site offers an online course for starting a new business. My Own Business Web site, http://myownbusiness.org/course_sba.html.

PlanWare. Helpful planning tips can be found at the PlanWare Web site, www.planware.org/strategicplan.htm.

State Chambers of Commerce. These organizations help "businesses do business." Their Web sites offer information about starting and maintaining a business in your state, including information on labor laws. For a link to the chamber in your state, see the U.S. Chamber Association Web site, http://www.uschamber.com/.

Software Programs

Microsoft Office. From version '98 on, the full version of Microsoft Office contains an interactive business plan writer that incorporates users' financial statements. It is available in the "Templates" folder.

Organizations Offering Business Plan Assistance

SCORE. This is a national association dedicated to helping small business owners form and grow their businesses. SCORE is a partner of the U.S. Small Business Administration (SBA). They have forms and templates to help you write a business plan, apply for a loan, or analyze your finances.

Small Business Development Centers (SBDCs). These centers provide management assistance to current and prospective small business owners. SBDCs offer individuals and small businesses a wide variety of information and guidance in easily accessible branch locations. The program is a cooperative effort of the private sector, the educational community, and federal, state, and local governments.

U.S. Small Business Administration (SBA). The SBA offers valuable Web resources and personal help to small business owners. It provides direct and guaranteed loan programs, and it co-sponsors the Small Business Development Centers that exist in conjunction with various states' Colleges and Economic Development agencies. For additional information, see the SBA.gov Web site, http://www.sba.gov/smallbusinessplanner/index.html.

Financial Resources

Many organizations help small businesses obtain funding. You should look into the following organizations and contact your county government as well. Most counties have an economic development expert who knows lenders in your county.

Center for Rural Affairs. Strategies for financing beginning farmers are offered here. For resources and loan programs for beginning and young farmers, see the Center for Rural Affairs Web site, http://www.cfra.org/files/BF-Financing-Strategies.pdf.

The Farm Credit System. This is a network of borrower-owned lending institutions and related service organizations that offer financial help to farmers across the United States and

Puerto Rico. For the contact in your area, check the Farm Credit Web site, www.farmcredit.com/.

USDA Farm Service Agency. In most counties, the Farm Service Agency is located within the office of the U.S. Department of Agriculture. The Farm Service Agency lends money and provides advice and credit counseling to farmers. It advises farmers and ranchers who are temporarily unable to obtain private commercial credit, such as beginning farmers who can't qualify for conventional loans because of insufficient net worth or established ranchers suffering financial setbacks from natural disasters. USDA Farm Service Agency Web site, http://www.fsa.usda.gov/ca.

USDA Rural Development. One of the key departments within the USDA is Rural Development, whose mission is to bring housing, modern telecommunications, safe drinking water, and a myriad other services to rural areas through its various loan and grant programs. USDA Rural Development Web site, http://www.rurdev.usda.gov/ca/.

REFERENCES

Covey, S. 1989. The seven habits of highly successful people. New York: Simon and Schuster Inc.

Krome, M., T. Maurer, and K. Weid. 2009. Building sustainable farms, ranches and communities. U.S. Department of Agriculture. National Sustainable Agriculture Information Service Web site, http://www.attra.org/guide/Building_Sustainable.pdf.

Rilla, E. 2011. Coming of age: The status of North Bay artisan cheesemaking. University of California Cooperative Extension, Novato, CA. UCCE Marin County Web site, http://cemarin.ucdavis.edu/files/73480.pdf.

Photo: Audrey Hitchcock

Parts of sections "Financing Your Business," "Points to Remember," and "Business Plan Resources" excerpted, with permission, from George, H. and E. Rilla. 2011. *Agritourism and Nature Tourism in California*. 2nd ed. Oakland: University of California Division of Agriculture and Natural Resources, Publication 3484.

Appendix 3.A. Business Plan Outline

Cover page

Table of contents

Executive summary

Financing proposal

  Desired terms

  Use of funds

  Collateral

  Owner's equity/cash contribution

Company description

  Brief history

  Description of products and services

  Description of your customers

  Business location and facilities

  Key strengths

  Owners and legal structure

  Planned changes

  Goals and objectives

Industry analysis

  Characteristics of industry

  Size of market

  Share of market

  Competition

  Barriers to entry

  Strengths, weaknesses, opportunities, and threats

Products and services

  List of products and services

  Distribution channels

  Competitive advantages

  Pricing structure

Market analysis

  Market demographics

  Product description

  Customers (characteristics, location)

  Competition (size, location, reputation)

  Location (requirements, accessibility, etc.)

  Marketing strategy (promotion, price, product placement, etc.)

  Market projections

Management and organization

  Descriptions

  Other structural necessities

Operational plan

  Methods of production

  Delivery

  Quality control

  Inventory control

  Credit policy

  Personnel

  Equipment, technology, and inventory

  Regulations and legal arrangements

  Exit strategy

Financial plan and projections

  History

  Plans

  Projections of production, sales, etc.

  Profit and loss

  Cash flow

  Projected balance sheet

Supporting documents

CHAPTER Four

Plant Layout and Design

Barbara Reed

Chapter Goals 32

Construction Materials 32

Sinks, Drains, and Plumbing Connections 33

Equipment, Ergonomics, and Sanitation 33

Layout 34

Environmental Controls 36

Brining Rooms and Cheese Aging Rooms 40

Planning for Growth 41

Points to Remember 42

References 42

CHAPTER GOALS

The goals of this chapter are to help prospective cheesemakers

- **understand basic concepts specific to dairy plant layout and plant sanitation**

- **consider worker ergonomics and safety**

- **consider special requirements for environmental controls, cheese aging, and energy costs**

- **plan ahead for growth**

If done poorly, the design and layout of your cheese plant can cause you lifelong headaches and backaches. You may also encounter significant problems with cheese aging and overall product quality. Doing things right the first time will pay you back many times over.

This chapter will point out the major factors that should be considered when designing your plant. These considerations go beyond the basic requirements of the Milk and Dairy Food Safety Branch (monitored by the California Department of Food and Agriculture, or CDFA), and the Food and Drug Administration (FDA)'s good manufacturing practices(GMPs). Work closely with your dairy inspector in this phase of your planning, as this person will provide feedback on materials and designs that are acceptable.

CONSTRUCTION MATERIALS

Dairy processing involves working with a raw product that is about 85 percent water. Once you make cheese and you concentrate the solids portion of milk (fat and protein), you still have to handle whey, allow cheese to drain, and use water for the cleaning and sanitation of the plant and equipment. Most of these processes will be done by hand and not in closed, mechanized systems. Your plant will have almost constant exposure to water and whey. Chemicals for cleaning dairy equipment include basic (caustic) and acid materials.

When building your plant, therefore, it will not be sufficient to use wood frame construction, put up sheetrock walls, and cover them with fiberglass-reinforced plastic (FRP) panels. These panels are commonly used in restaurants and food manufacturing facilities because they are an impermeable surface that can be cleaned easily. Older, thin FRP panels are great, but they are more appropriate for a facility like a bakery, which has limited water exposure and low humidity. If damaged, the thin FRP panels no longer provide a surface impermeable to dirt, moisture, and pests. This type of FRP panel cannot meet the demands of cheese plant manufacturing. However, newer structural FRP panels with foam cores have been developed, and these are more appropriate for cheese facilities.

If you consult with your architect and decide not to use the newer FRP panels for wall construction, you will need to use ceramic tile applied to a cementitious, water-resistant board (like Hardibacker) or poured cement that has been sealed with an acceptable coating (such as epoxy coatings). Cement blocks can be used as long as all openings are sealed, blocks are well aligned, joints are flush (so they do not accumulate dirt or debris), and the surface is sealed.

"Visit as many cheesemaking facilities as you can and take notes. The most frequent mistake I see—and we made it, too—is planning for too small a space or cheese operation. You will need a certain amount of sales to cover your costs. Be sure you have the space to make the cheese you need to sell. Decide ahead of time what you need to sell, based on what your costs will be. Be sure to plan for employees. Cheesemaking is labor intensive and even the smallest cheese operations have some outside labor. With that you need to plan for lockers and break room, workman's compensation insurance, and training."

Cheesemaker Story

Work with your dairy inspector and obtain the professional help of an architect and construction company familiar with cheese-manufacturing requirements. Be sure that both the walls and the floor are properly sealed and insulated so you won't have unwanted condensation occurring on these surfaces. This is especially important in aging rooms where the humidity must be carefully controlled.

If construction, insulation, ventilation, and drainage are inadequate, the structural elements of your building may rot or uncontrollable mold growth could develop. In a worst-case scenario, *Listeria monoctyogenes* could take up residence in your plant and contaminate your product. Unless the plant were demolished and rebuilt, it could never return to production.

Floor and wall junctions are critical. The intersection of walls and floors must have a smooth, seamless junction (coved or rounded) that allows for adequate cleaning and prevents entry of pests. The easiest way to accomplish this with cement materials is to have the walls and floors poured so they form a single, contiguous surface. With tile walls, coved tiles are used at the floor and wall interface or cement coving can be formed after the walls and floors are complete. Inspectors may require that the flooring be sawn so that the cement coving "locks" into the floor when it is poured.

SINKS, DRAINS, AND PLUMBING CONNECTIONS

Sinks, drains, and plumbing connections (including hoses) will be a major focus for you and your inspector. Plant design must ensure that there is no cross-connection between a safe water supply and any unsafe water supply. The design should also protect the water supply from contamination by chemicals or food residues like whey, etc. This means there must be air gaps and backflow prevention devices installed at critical points in your plumbing system to protect against contamination. In addition to faucets and other water connections, you need to have proper hose location, connection, storage, and use. In food processing areas, plan to have hoses on reels that will keep them off the floor, especially if hoses are used in or around the cheese vats.

Floors need to be properly sloped to drains, and drains should be constructed so that they can be cleaned. The slope needs to be about 1 inch per 10 feet. Trap drains are preferred over trench drains because they are easier to construct and clean. If trench drains are used, they need to be sloped so that there is never standing water in them. This is hard to accomplish if the floor is wet most of the time.

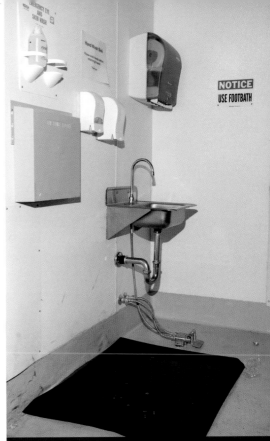
Hand-washing stations that operate with sensors or foot pedals should be located in the processing area. *Photo: Mike Poe.*

Hand-washing sinks should have hands-free faucets and be operated by foot pedal or electric eye, and the same goes for paper towel dispensers. Don't skimp on sink placement and number of hand-washing stations. Convenient and frequent hand washing is one key to good manufacturing practices (GMPs).

EQUIPMENT, ERGONOMICS, AND SANITATION

According to the U.S. Department of Labor, food manufacturing has one of the highest incidences of injury and illness among all industries. Cheesemaking is physically taxing, repetitive work. Create a plant set-up that will reduce back strain, repetitive motion,

Repeatedly bending over a deep vat is hard on you and your employees. Consider ergonomics in your plant design to reduce repetitive motions that can cause fatigue or injury. *Photo: Barbara Reed.*

and the risk of slips and falls. Design considerations should include the depth and height of the cheese vat, accommodation for gravity flow of curds, and equipment placement and installation for ease of cleaning. If you can set up elevated cheese vats and move the curd into cheese molds by gravity rather than bending and lifting, everyone will benefit.

Include enough floor space in your layout so that personnel can work around and clean all pieces of equipment. Incorporate movable equipment into the plan so that workers don't have to adopt awkward postures to use or clean equipment. Remember, good manufacturing practices require a clear perimeter around the edges of work space for the sake of cleaning and pest detection.

LAYOUT

The easiest way to keep your plant clean is to not get it dirty in the first place. Lay out the plant so that product flow is linear, and so that personnel and finished product do not move through the same spaces as raw product or product in the midst of processing. If work space is set up properly, there will be no need for personnel other than the cheesemakers to be in the manufacturing area. Provide separate work areas for cheese that is in presses or draining in molds so that this cheese does not remain in the manufacturing area when cleaning begins.

To prevent pests and dirt from getting in, the manufacturing area should not have doors or windows that open directly to the outside, and the number of entrances and exits to the manufacturing area should be limited. If possible, manufacturing, brining, aging, and packaging areas should have positive air pressure. Hallways and breezeways should be incorporated into the design to prevent direct access from the outdoors into processing rooms and so that the processing room does not serve as a hallway. Footbaths should be placed at entrances to the plant and processing areas. If personnel change into work clothing at the plant, provide a separate changing area for them to do this. The area may be adjacent to but should be separate from

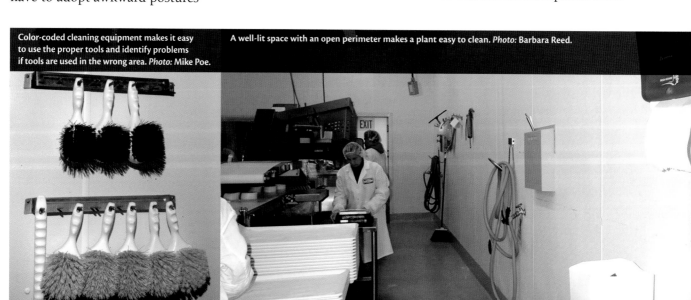

Color-coded cleaning equipment makes it easy to use the proper tools and identify problems if tools are used in the wrong area. *Photo: Mike Poe.*

A well-lit space with an open perimeter makes a plant easy to clean. *Photo: Barbara Reed.*

Figure 4.1. Ground floor plan for a cheese plant.

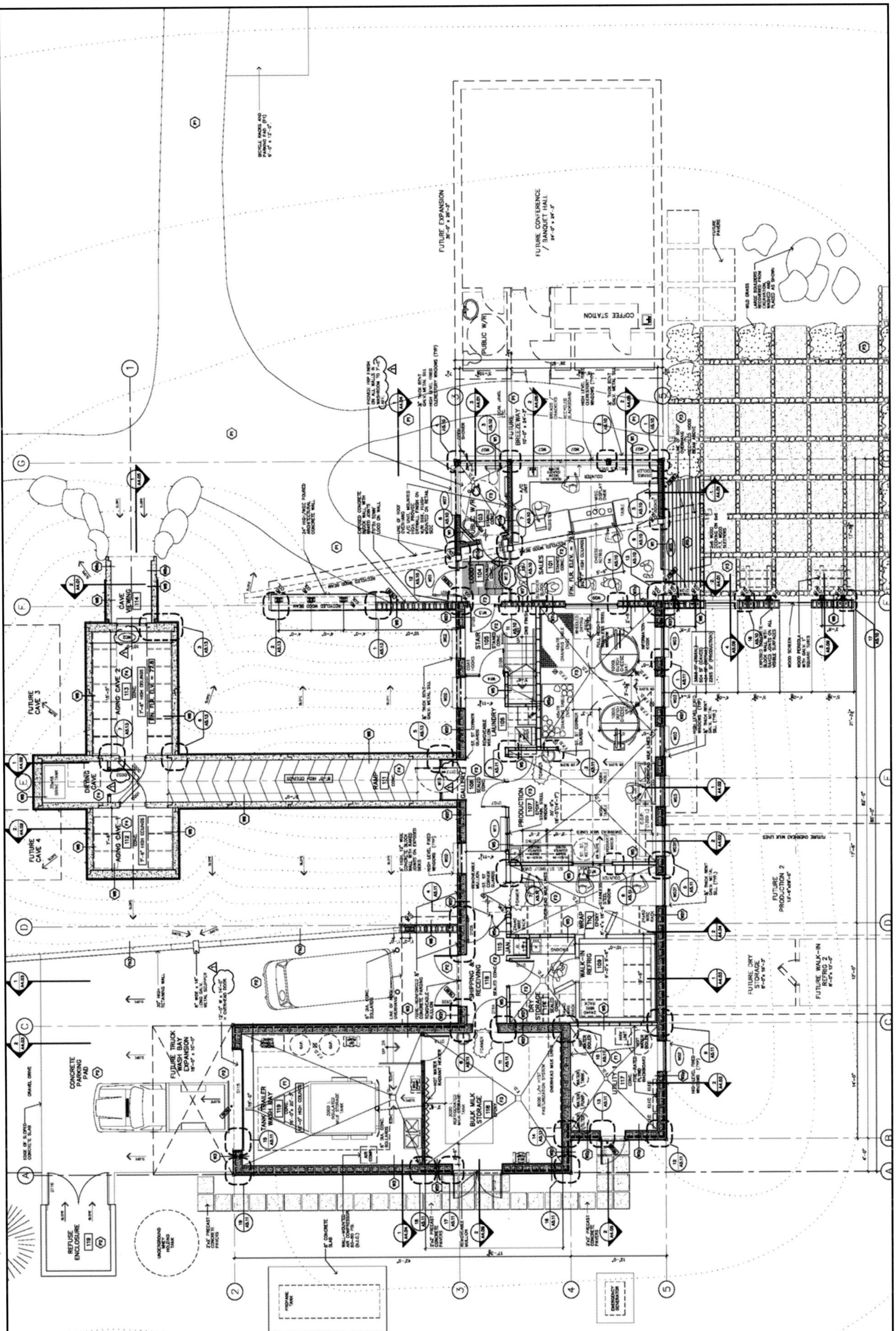

Source: Courtesy of LaPointe Architects.

bathrooms, because you don't want to be storing uniforms and other clean materials in the bathroom.

Figures 4.1 to 4.3 are included to show the complexity of a well-planned cheesemaking facility. They are not intended to provide a template for design or to imply that one particular layout is appropriate for all cheesemaking. The movement of personnel and the product flow is highlighted in figure 4.2. Figure 4.3 provides a further detail of equipment layout for operator comfort and work flow. Note in figure 4.3 how the people are included in the overall design and layout of the work space and equipment placement. Highly simplified examples of product flow are shown in figures 4.4 and 4.5. Figure 4.4 describes product flow by process while figure 4.5 shows a graphic representation of product flow direction.

ENVIRONMENTAL CONTROLS

Controlling energy costs will be an important element of your profitability. You can reduce the cost of plant operations by incorporating passive design features like daylighting for work space, adequate insulation, heat exchange/recirculation,

Figure 4.2. Product and people flow diagram (detail).

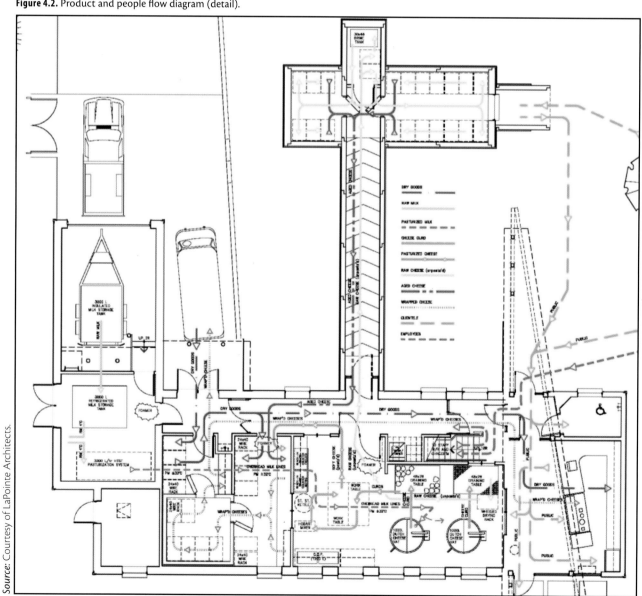

Source: Courtesy of LaPointe Architects.

Figure 4.3. Ground floor plan enlargement.

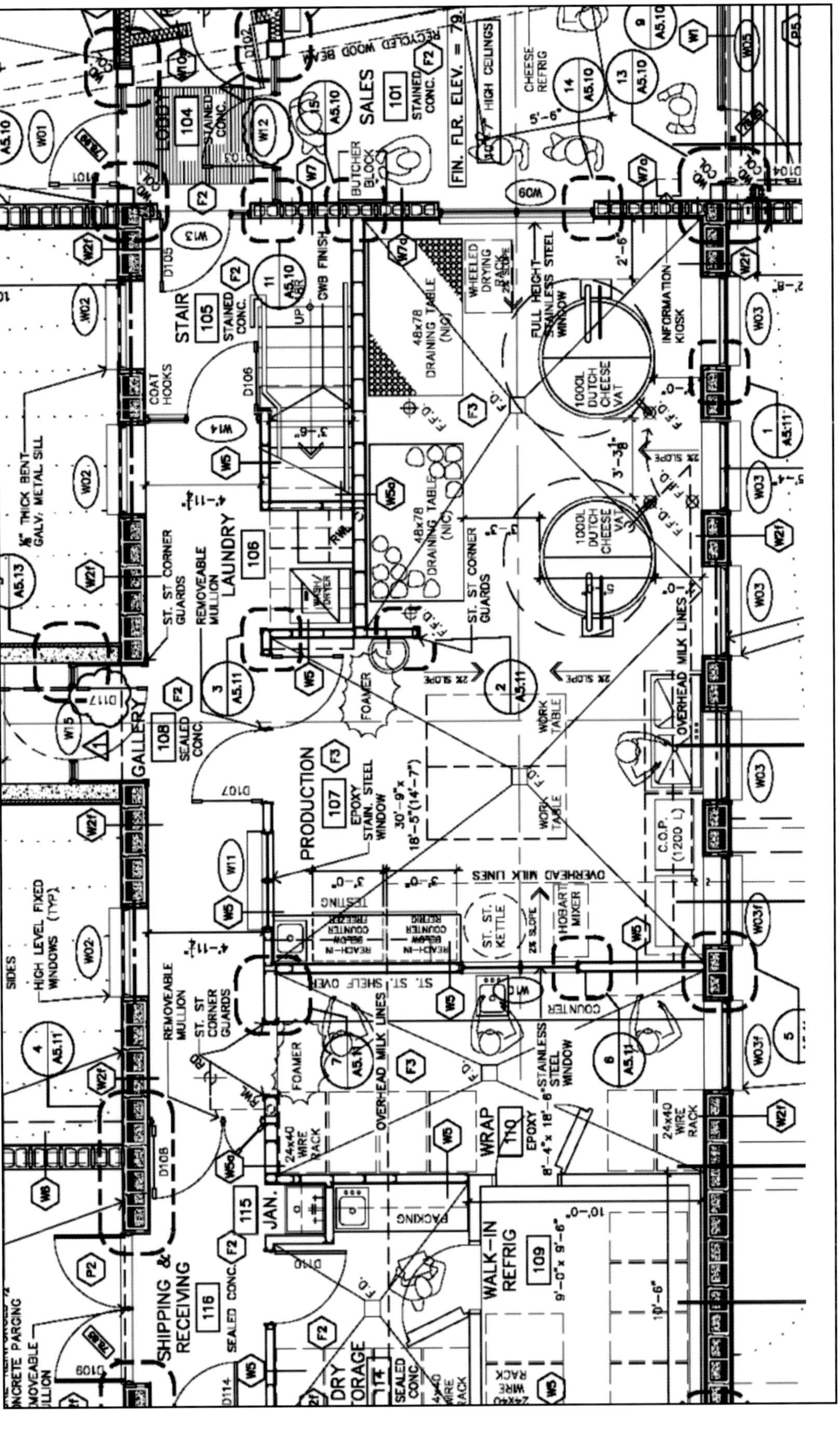

Source: Courtesy of LaPointe Architects.

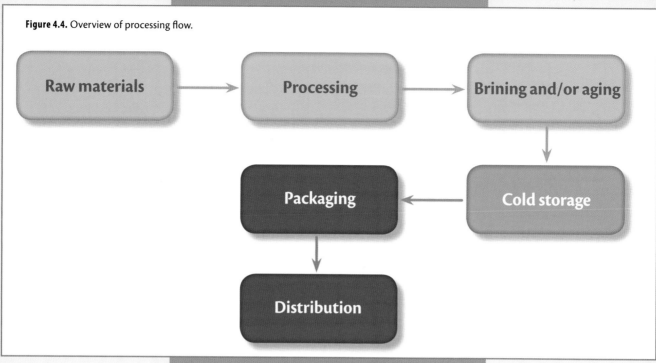

Figure 4.4. Overview of processing flow.

Raw materials → Processing → Brining and/or aging

Packaging ← Cold storage

Packaging → Distribution

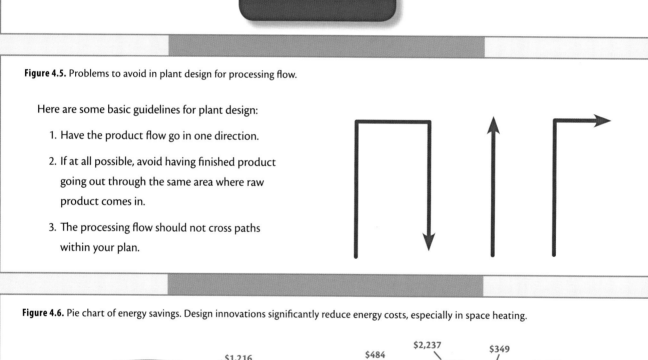

Figure 4.5. Problems to avoid in plant design for processing flow.

Here are some basic guidelines for plant design:

1. Have the product flow go in one direction.

2. If at all possible, avoid having finished product going out through the same area where raw product comes in.

3. The processing flow should not cross paths within your plan.

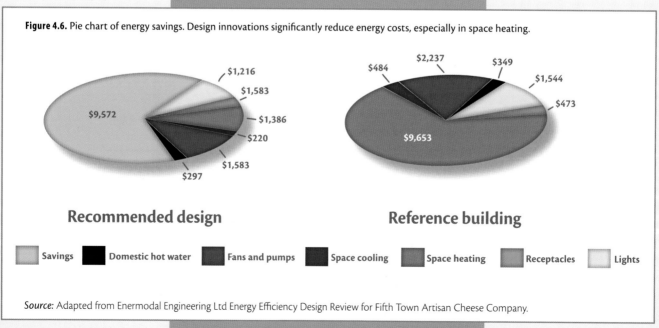

Figure 4.6. Pie chart of energy savings. Design innovations significantly reduce energy costs, especially in space heating.

Recommended design — $9,572, $1,216, $1,583, $1,386, $220, $1,583, $297

Reference building — $484, $2,237, $349, $1,544, $473, $9,653

Savings | Domestic hot water | Fans and pumps | Space cooling | Space heating | Receptacles | Lights

Source: Adapted from Enermodal Engineering Ltd Energy Efficiency Design Review for Fifth Town Artisan Cheese Company.

and water conservation. You can help minimize the environmental impact of your facility by working towards Leadership in Energy and Environmental Design (LEED) certification. This is a rating system developed by the U.S. Green Building Council (USGBC) to assess the environmental sustainability of building designs. Work with your architect, dairy equipment professional, and your local Heating, Ventilation, Air Conditioning (HVAC) contractor to calculate water, heating, and cooling needs for your plant. Do not guess at this. If you skip these calculations, you can end up without enough gas supply to run your hot water heaters or boiler. Furthermore, you may end up wasting energy, and these costs will erode your profit margin.

You can also hire an engineering firm that can provide costing databases and simulation tools to identify cost-effective approaches to get LEED credit. These engineering firms can perform energy and daylighting simulations, and they can design the mechanical and electrical systems in conjunction with your local tradespeople. Refer to table 4.1 to see the potential benefits of energy planning.

Tank Sizing for Batch Pasteurization versus High-Temperature Short-Time (HTST) Pasteurization Equipment

As long as your production volume remains below 550 gallons, you should probably stick with batch pasteurization. HTST systems cost between $75,000 and $100,000, and they are separate from a cheese vat and other processing equipment. A batch pasteurization/cheese vat combination can be fabricated for about $25,000 and can handle up to 550 gallons.

Once you get beyond a volume of about 550 gallons, too much energy and time are required to heat, hold, and cool a large volume of milk. When the volume of milk becomes too large, there is no net energy or cost saving with batch pasteurization. The number of BTUs needed to heat milk can be derived from the following equation:

$$BTU = \text{pounds of milk} \times \text{each increase per } °F$$

Be sure you understand your equipment needs so that you don't buy equipment you don't need. Don't invest in a steam boiler if all your heating needs for food processing can be done with hot water. With assistance, you can calculate the relative energy costs of a small, high-temperature short-time pasteurizer, rather than using a batch pasteurizer.

To help conserve valuable resources, be sure to add a tank to store water from your heat-exchange chiller. You can set up your system so that pasteurization vats can utilize this preheated water.

The items mentioned here touch on the mechanical and electrical systems that will be installed in your plant. When designing your plant, have your local equipment contractor team up with a company that specializes in energy conservation. Figure 4.6 shows how one plant

Filling small forms properly can't be hurried. *Photo: Steve Quirt.*

Producing good cheeses can be a labor-intensive process. *Photo: Steve Quirt.*

Choosing the right vat size and shape is critical. Combining pasteurizing and cheesemaking functions may be important to your operations. This vat is not set up to pasteurize milk. *Photo: Steve Quirt.*

Table 4.1. Reducing long-term energy use and associated costs in cheese plant operation*

Type of design	Anticipated energy performance			Savings
	Electricity cost	Natural gas cost	Total energy cost	
Reference building	$13,807	$933	$14,740	—
Proposed design	$4,259	$909	$5,168	62.60%

Source: Adapted from Enermodal Engineering Ltd Energy Efficiency Design Review for Fifth Town Artisan Cheese Company.

Note: *By analyzing energy use and identifying ways to save energy through innovative design, materials, and engineering, this cheese plant was able to identify a two-thirds reduction in energy costs over conventional building design.

was able to cut its expected energy costs by two-thirds as a result of using innovative energy-saving strategies.

BRINING ROOMS AND CHEESE AGING ROOMS

You cannot make three different kinds of cheese, put them all into the same walk-in refrigerator, and hope that it will all work out. A commercial walk-in cold box with a big fan moving lots of air around may be perfectly fine for refrigerating your brine vat or for boxed cheeses awaiting transport,

but it is a marginal choice for a cheese aging room. As table 4.2 illustrates, once the cheese leaves the vat or the brine tank, it may need to be placed in a series of controlled environments before it is ready for sale.

You need to design these environments based on predicted rolling volume and cheese process requirements. Shelving materials may need to absorb moisture or allow for air circulation. Different cheeses may not age together because molds of one cheese could contaminate another. Too much air movement may

desiccate cheeses. In cheese aging research, experts have found that the days to optimal development of a ripened cheese changes significantly when the aging room environment changes by just one degree! If you have to store cheeses longer than necessary, inventory turnover and cash flow slow down, and the energy cost per unit of cheese is greater. Additional cheese moisture is lost with longer storage times. Part of the cheese weight you sell is water, and you will lose income with excessive moisture loss, also called shrinkage. If the humidity control is way off, you may end up with wheels of cheese that are cracking because they are too dry. And with these severe defects, they may be entirely unsalable. If humidity is too high, cheese may be contaminated from condensate forming on refrigeration units, walls, and ceilings, and then dripping onto cheese. You should design your facility with enough space so that aging rooms can be emptied completely for cleaning.

According to the experts at the University of Guelph's

Storing cheese for aging requires good space planning. *Photo: Steve Quirt.*

Table 4.2. Temperature and humidity parameters for maturation of various cheeses

Cheese	Process	Temperature range (°F)	Relative humidity (%)	Aging time (range)
Asiago	maturation	50–54	85–90	4–18 months
Brie	drying	54–58	ambient	6–7 days
	mold formation	50–58	80–85	until white mold develops
	maturation	47–50	80–85	4–5 months (until ripe)
Cheddar	drying	ambient	--	--
	maturation	45–52	ambient	until ripe
Emmental	drying	50–61	90	10–14
	warm maturation	68–76	80–85	--
	cold maturation	45	85–90	--
Gouda	drying	59	80	6–7 days
	storage	50	80	4–6 weeks
	maturation	50	80	6–12 months
Parmesan	--	50 maximum	below 75	--
	--	41–50	below 75	2–4 years
Roquefort	cooling	47–50	95	18–25 days
	maturation	41–50	95	not specified
	final ripening	34	95	5–10 months

Source: Adapted from Scott, Robinson, and Wilbey 1998.

Department of Food Science, most European varieties are stored at 50° to 59°F (10° to 15°C) for initial ripening and then at 39°F (4°C) until consumed. Surface-ripened varieties are ripened at 52° to 59°F (11° to 15°C). Surface-ripened cheeses also require adequate air circulation to provide sufficient oxygen for molds and yeasts. Relative humidity (RH) requirements in general are

- washed, bacterial, surface-ripened: 90 to 95 percent RH

- fungal flora: 85 to 90 percent RH

- dry rinds: 80 to 85 percent RH

PLANNING FOR GROWTH

Planning for growth is essential in food-manufacturing operations. You can't just shut down production and start up a year later when your new (larger) plant is finally done. There are some things you can do in your initial planning phase to make your inevitable expansion go smoothly.

Brine needs to be meticulously maintained to be sure that the solution is saturated and particulate matter filtered out as needed. Brines can become contaminated if not cared for properly. *Photo: Steve Quirt.*

When you lay out your plant in the beginning, also figure out how you will add on to the facility. If possible, design process flow so that when expansion comes, you duplicate your processing rather than scaling it up. In other words, add a second cheese vat rather than buying a bigger cheese vat. When you duplicate processing rather than scaling up, you don't have to adjust all the formulations in your manufacturing process; instead, you will just have to manage the logistics of repeating the process. Work with your equipment experts to determine what segment of your processing will be the limiting factor that forces expansion. For example, you may need to add another brining tank and more aging space even before you have to purchase a second vat.

POINTS TO REMEMBER

↠ Include plant expansion considerations in your initial planning.

↠ Duplicate processes rather than scaling up.

↠ Identify what part of process flow is the limiting factor.

REFERENCE

Scott, R., R. K. Robinson, and R. A. Wilbey. 1998. Cheesemaking practice. New York: Kluwer Academic/ Plenum Publishers.

Adding a second vat to duplicate processing is simpler than having to scale up your make process for a larger vat. *Photo: Valley Ford Cheese Company.*

Sections of Chapter Four on cheese plant construction adapted in part from Schmidt, R. H. and D. J. Erickson. 2005. Sanitary Design and Construction of Food Processing and Handling Facilities. Gainesville, FL: University of Florida, Institute of Food and Agricultural Sciences, Florida Cooperative Extension Service, Food Science and Human Nutrition Department series, Publication FSHN04-08. University of Florida IFAS Extension Web site, http://edis.ifas.ufl.edu/fs120.

The sections "Layout" and "Planning for Growth" adapted in part from University of California Cooperative Extension Farmstead Cheesemaking Workshops, 2003–2005. Glenn County Cooperative Extension, Orland, CA.

The section "Brining Rooms and Cheese Aging Rooms" adapted in part from Food Safety Training for Cheesemakers: Prerequisite for HACCP Workshops, 2004–2005. Glenn County Cooperative Extension, Orland, CA.

CHAPTER *Five*

Designing Your Marketing Strategy

*Leslie J. Butler, Barbara Reed,
Ellie Rilla, and Holly George*

Chapter Goals 46

Introduction to Distribution and Margins in Specialty Cheese 46

Distribution Methods 46

Margins 52

Freight 53

Marketing Planning for a Small Cheese Plant 54

Points to Remember 77

References 77

INTRODUCTION TO DISTRIBUTION AND MARGINS IN SPECIALTY CHEESE

Before you begin your marketing plan, study the information about specialty cheese sales and distribution presented in this chapter. Although sales and distribution will be a part of your total marketing plan, it will be important to understand these details first, so that you can develop your marketing strategy around the unique constraints of specialty cheese sales and distribution.

As the proprietor of a small business, it will be important to maintain control of your product, pricing, and distribution. Make sure that the marketing channels you use work to your advantage and meet your business objectives.

DISTRIBUTION METHODS

Channel strategy. This strategy lays out the distribution methods and partners you will use to reach customers in the target markets you have selected. (We will discuss your target market a bit later.)

Direct sales

This is the most straightforward type of marketing—directly from producer to consumer. In the strictest sense, direct sales are from you (through your booth at the farmers' market, your farm store, or your Web site) directly to the consumer. In a broader sense, direct sales can be made to a reseller like a retail grocery store or high-end restaurant, but without a broker or distributor as middleman. The goal of direct sales is to capture additional revenue that otherwise goes into distributor and/or retail margins. (We will talk more about margins later.) The cost of direct sales will include your labor (including your managerial effort), expenses for hired sales staff, permits for farmers' markets, cost of booth space, Web hosting, packaging and transporting product to market, and parcel shipment costs (packaging, cold packs, labor, sales inserts, invoices) for Internet sales.

The main limitation of direct sales not done through Internet is that you sell in small quantities in a limited geographic area. However, this is an ideal quality-assurance mechanism for a perishable product. When you first get started, direct sales allow you to build your market slowly, maintain control over where your product is sold and how it is handled prior to sale, and retain more sales dollars.

"Our biggest challenge was sales and marketing, learning new skills that we don't need as dairy farmers."

Cheesemaker Story

Cheese brokers

Most brokers never take ownership of your cheese, although they represent you and your product in the marketplace. They will charge you a commission of between 5 and 10 percent on sales and may charge a monthly fee as well. Hiring a broker is a good idea if you want to move from local to regional marketing. Brokers can be very useful in extending a small sales force. They can be appropriate for hard-to-reach areas where you may still be able to manage your own shipping but may not have time to talk with all the customers who might buy a small volume of your product. The broker will also represent other cheeses (and perhaps other specialty foods) and be able to leverage sales by promoting a wider variety of products to the customer than you can as an individual vendor. If you work with a good broker, he or she will be extremely knowledgeable about specialty cheeses (including yours), understand the best market fit for your product, have a good clientele base already established, and work hard to move your product. When introducing your cheeses through a broker, you may want to start with several products. If you have only one product, it limits your potential volume to retailers and may make it harder to develop interest in your brand. You can find a broker through recommendations from your peer cheesemakers, through organizations such as the American Cheese Society, or by speaking with cheesemakers at specialty food events like the Fancy Food Show held on both coasts.

Distributors

When choosing a distributor, you want to find one that focuses on specialty food and sells in your target market. This is often the most efficient way of distributing your cheese when products are sold in different retail settings all over

the country. Distributors take physical possession of your product. It is best if you have experience with the grocery market before you start using distributors and selling to larger grocery chains. Distributors buy in volume and resell in case lots. You do not want to be in a distribution system that sells bubble gum, chewing tobacco, and nacho cheese to a convenience store. Specialty food distributors may carry other fine food items such as olives, chocolates, and imported meats and cheeses. Go to stores that carry specialty cheeses comparable to your product and find out who distributes those cheeses. Interview those distributors, ask them what areas they serve, what stores they sell to, and what other cheeses they sell. Check out several retail stores and see how those products are featured and how they are priced.

Your distributor will purchase your product outright at a price referred to as freight on board (FOB). In theory, distributors will take possession of the cheese as soon as it gets loaded

Specialty cheese departments at retail grocery stores are now commonplace. *Photo: Barbara Reed.*

When doing your market research, see how back-stocked cheeses are stored. They should be well cared for like these cheeses. *Photo: Barbara Reed.*

onto *their* truck at *your* loading dock. They pay you on terms that have been mutually agreed upon, such as "30 days same as cash" or 10 percent discounts on whole-pallet orders. Then you kiss your cheese goodbye and wish it well.

While the FOB arrangement works as described in most distribution systems, it is more complicated for small cheese producers and those whose plants are located in rural areas far from main truck routes. Refrigerated trucking and storage is critical for maintaining product quality until point of sale. Find out if you are even on an existing route for distributors and if they have a minimum pick-up volume (e.g., a pallet or more).

Depending on the cheeses (large wheels versus small wheels, or vacuum-packed random weights), a full pallet may weigh anywhere from 800 to 1,400 pounds. As a small-scale cheese producer, you may have less than a pallet of cheese to move at any one time. And those

A distributor won't send an 18-wheeler off a main trucking route to pick up 500 pounds of cheese. *Photo: Mike Poe.*

cheeses may need to go to eight different retailers. A distribution company is not going to send an 18-wheeler off a major highway to pick up 500 pounds of cheese. In fact, distributors may choose not to work with you if your production volume is too small, or at the very least they will require you to arrange (and pay for) product shipment to one of their central distribution points before they consider it FOB.

If the distributor company requires you to get your product to its distribution center, this may take some ingenuity. You may be able to find some other freight company (produce, dry goods) that delivers to your area and is willing to do backhauls (think food safety first!) when the truck is empty. If that is not an option, you may have to make your own deliveries to the distributor, or ship your cheese along with your Aunt Louise when she "goes down to the city" once a month. Whether you use backhauls or Aunt Louise, it is critical that food safety is considered and pre-FOB costs are factored in to your "FOB price," or you will be losing money and cheese before your cheese reaches a consumer.

Make sure that you carefully review your contract with a distributor. You should have payment terms specified, such as "2 percent, 10 net 30," which means that a 2 percent discount is provided if payment is received within 10 days of

the delivery of goods, and that full payment is expected within 30 days. Be clear about "charge backs"; distributors may bill you by deducting from your invoice for demos, ads, slotting fees, damaged or out-of-code product, and trade shows. When a distributor company agrees to work with you, it may ask for an exclusive contract for a certain period of time. This means you could not work with another distributor or broker that serves the same market area until the exclusive expires. This avoids having multiple distributors in one area competing to sell the same product. One limitation of exclusives is that your distributor may sell to retailers, but not to foodservice. So working with another distributor in the same area might make sense if you want to sell into a part of the market not served by the first distributor.

Because specialty distributors work with a large product list, their sales representative will not be able to promote your product as well as you or your broker. It will be important to provide the distributor with point-of-sale materials and take time to educate that company's sales staff about your product.

When introducing a new product, the distributor may ask you to "demo" your cheese at a retailer. This means in-store sampling, which is an important opportunity for you to tell your story directly to the consumer and drive your own sales. It also

gives you a chance to meet some of the folks at retail who will be caring for and displaying your cheese. Demos are important in any sales setting, whether it is a farmers' market, a trade show, or a retail setting. Along with samples, you are able to share point-of-sale materials and make a personal connection with consumers.

Types of distribution

- **Exclusive distribution:** Product is available only at exclusive outlets, which builds customer loyalty and allows the retailer to charge higher markups. Market is limited, but this makes distribution easier because it cuts down on the number of outlets.

- **Selective distribution:** Product is not available to every retailer but to selected retailers who meet your sales specifications. Restricted availability allows the retailer to increase volume sales without competition from other local retailers.

- **Multiple-channel distribution:** This may be needed if you plan to be in multiple geographic areas even with small volumes of product. Many larger firms use multiple channels to distribute their product. A single specialty cheesemaker may use as many as 40 distributors to reach his or her customer base.

As you can see in figure 5.1, most distributors are concentrated in urban areas to be close to their target markets and close to major trucking routes to facilitate shipping logistics. One complex aspect of working with distributors is that they will not necessarily share with you a complete sales history of your cheeses, nor tell you what their wholesale price is to the retailer. The better ones will give you a sales report, even if they don't share their revenue information. Generally, the distributor's margin will be around 23 to 25 percent. Keep in mind that distributors will be sure to recoup any of their freight costs to get your product to their customer, and so will calculate their cost of freight into the margins. In other words, they will make money on their freight costs.

Distributors will also negotiate promotional deals with you. They may want a certain amount of free product for sampling and demos, especially when going into a new market, or they may ask for a discounted FOB price so they can run sales promotions on a product line. You need to build all of these elements into your marketing costs so that you don't give away your inventory!

Retail

The previous brief discussion about in-store demos gives us a good chance to talk about retailers. Even if you are working with a distributor, it is helpful for you to understand retail sales operations. Most people do their food shopping in retail stores, and so retail grocery stores provide a large part of the potential market. However, for most small-scale cheesemakers, there are special challenges in the retail sector. Most large grocery chains won't carry your product until you are an approved vendor. First, you have to provide product and production information to their regional or national buyer. The buyer may or may not want you present when they sample your product and make a decision about carrying your cheese. Once you have been approved, you will be on their approved vendor list. At this point, you may be able to work directly with the local store, but more likely you will have to work with one of their authorized distributors.

A smaller specialty grocery store close to you is a great way to learn about the grocery channel. You may even work directly with the store's cheese buyer. This person is usually part of the deli department, if the store doesn't have the staff for an entire cheese department. When you have an opportunity to work one-on-one with these departments, it will be important for you to educate them about your cheese, including information on your farm and

Figure 5.1. Nationwide map of some distributors' headquarters used by California artisan cheesemakers.

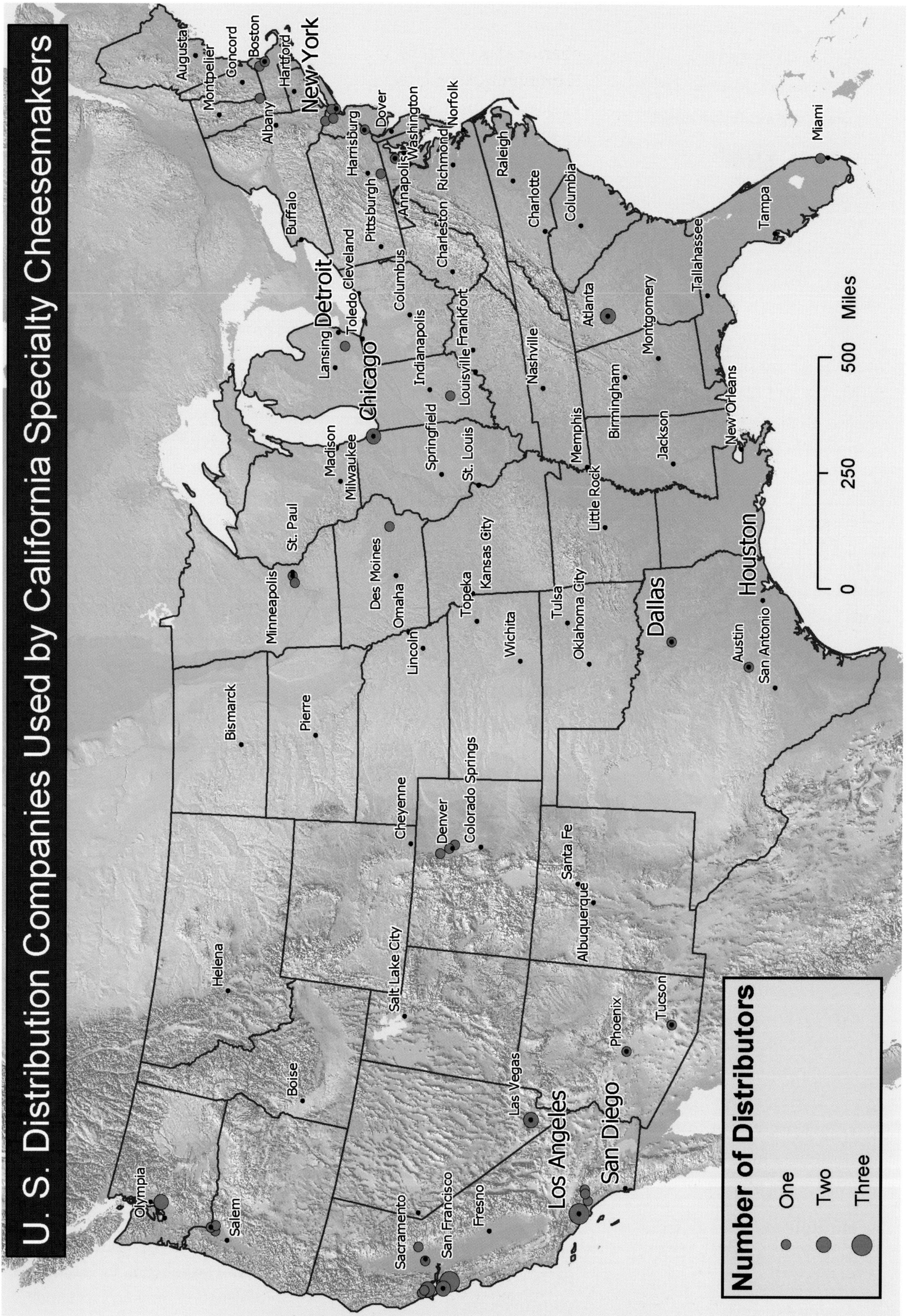

family, attributes of the cheese, and suggested food and wine pairings. They will also educate you about their marketing practices and consumers. Consumers love to hear stories about their specialty cheese. Again, it will be important to participate in demos if given the opportunity. Provide the store with plenty of point-of-sale materials and be sure it has a high-quality digital photo of your cheeses so it can feature them in its newsletter.

When trying to make a decision about which retail stores you may want to sell through, take time to look at their cheese displays, see how the staff interacts with the customers, and how much time they take to make a sale. If you are in a store that provides samples, observe how the employees serve customers, what information they convey to the consumer, and how many samples they offer. Look at the display cases and how the cheeses are presented. Notice if they look fresh and appetizing, or if they look old and as if they've been handled improperly. Have the display cases been rummaged through, and is it hard to find a product you are looking for? What information is provided on "shelf talkers" (also known as case cards) or package labels? Is there a QR (quick response) code that a consumer can read with their smartphone, linking them to detailed information about the cheese? Is information about food and beverage pairing available? Is the cheese case full service and

offering cheeses cut to order, or is everything prepackaged?

Good specialty grocer employees will offer a lot of cheese samples for consumers to taste, give them plenty of time to make their selections, and be very knowledgeable about their cheese inventory. In larger retail settings like Safeway, specialty cheeses are usually prepackaged and in their own case, and only occasionally available for sampling.

The retail margin depends on whether you are dealing with Costco, Safeway, a regional grocery store, or Dean and Deluca. Costco may have a smaller margin than Dean and Deluca (good for the consumer), but they can work with those margins because the contract they negotiate with you guarantees that you will ship a certain volume of product over time and accept a deeply discounted FOB price. Safeway or Dean and Deluca may have much higher margins but won't necessarily have such high volume or discount requirements. Major grocery chains may require other incentives in order to place your product on their shelves. Both

distributors and retail stores may have requirements related to food safety and quality assurance. Some of these requirements might be the following:

- Ship product in a refrigerated carrier approved by the retailer (no UPS or FedEx shipments, only specific distributors).

- Provide evidence of product liability insurance.

- Have third-party Hazard Analysis and Critical Control Point (HACCP) certification (or that of another quality assurance program) of your manufacturing facility.

- Make sure that lot and date of manufacture, along with correct cheese names and weights, are visible on stacked boxes. Boxes should contain enough repack labels for random-weight packaging at the store if you sell whole wheels.

Figure 5.2 provides some real-life market feedback about marketing artisan cheese to retailers.

Figure 5.2. Market considerations from retail field research.

- European cheeses pose serious competition to American artisan cheeses on the East Coast. They are price competitive, and consolidation and cooperative marketing have already been worked out with fellow European cheesemakers. In addition, European dairy production has financial support from European Union policy. Are you competing with European cheeses?

- Any new packaging techniques should be tested locally before shipping to a new, distant market.

- How you create samples for distributors and potential new retailers (in terms of size, shape, and ability to demonstrate visual characteristics of larger wheels) is critical to effective sales efforts.

Foodservice

Because you are making a specialty product, foodservice sales should focus on high-end restaurants or bakeries that will value your product over a generic cheese and communicate this value to their clientele. The margins added in foodservice depend on whether the item is used as an ingredient or featured as a course. If used as an ingredient, the cost of the dish will be calculated in total and then the overhead of costs of labor, etc., are added on, but the final consumer will not see a price for your cheese. If your cheese is used as a course, the markup would be similar to wine and could run upwards of 200 percent. Foodservice sales may require some adjustments to your operation. If your cheese is an ingredient in desserts or other dishes, you may be asked to provide product in bulk. For example, if you make a quark to be used in a cheesecake, the restaurant may want it in 5-pound tubs. The advantage of selling into foodservice is that they may be able to order larger quantities than if you sell exclusively to retail. However, you may have to discount the price for bulk sales and use additional packaging and product-handling methods in your plant to accommodate these customers.

Master distributor

You can think of the master distributor as the middleman's middleman. Let's say you want to ship cheese to the East Coast. You ship cheese to a master distributor in New Jersey that sells to other regional distributors all along the Eastern Seaboard. This means that additional price increases will affect your product before it gets to the consumer. The same is true if you are shipping to the West Coast to a major distributor.

Generally, you should avoid working with master distributors because the added margin will move your retail cheese price substantially higher. The retail price in the markets directly affects the sales volume. The higher the price, the lower the sales. Because of the way distribution channels are set up, the producer has little control over retail price when too many middlemen are in place.

MARGINS

As mentioned in the introduction, current market data indicates that the market will grow and plenty of room exists for the new cheese that you have in mind. Specialty cheese is a high-end food that implies big profits from a sophisticated clientele. However, many cheesemakers have started out making simple (but incorrect) calculations of their future profits. Here is an example of their wrong assumptions: "If I can purchase raw cow's milk for about $15 per 100 pounds of milk, and it takes about 10 pounds of milk to make 1 pound of cheese, then I can make about 10 pounds of cheese for just $15. If my artisan cheese sells for about $18 per pound, then that would be a margin of about 1,200 percent!! Well, okay, I have to pay for things like establishing the cheese plant, rennet and salt, energy to make it, and, of course, pay the labor, too, and maybe something to market the cheese. So even if that halves my margin to about 600 percent, that's a pretty good return on my money."

People unfamiliar with the cheese business assume erroneously that they will capture 90 to 100 percent of retail cheese value. The realities of the cheese business are really quite different. First, unless marketed directly, the cheesemaker does not receive the retail price for the cheese. Because you are using a distributor and retailer to market your cheese, 70 to 80 percent of the retail price of all food remains within the marketing sector. This is due to their margins, which are needed in order to cover their costs and allow them some profit. In other words, most cheesemakers receive about 30 percent of the retail price of their cheese when sold through distribution and retail channels.

A margin works like this: You have a price of $5 per pound for your cheese when selling to a distributor. Your distributor adds its company's margin of 23 percent.

Distributor price ÷ (1.00 − margin percentage) = price to retailer
$5.00 ÷ (1.00 − 0.23) = $6.49

A retailer then adds his or her margin to that figure:
Price to retailer ÷ (1.00 − 0.60) = $16.25

So $16.25 would be the price that the consumer pays for the cheese.

Distributor's margin:
23 to 25 percent

Master distributor's margin:
8 to 12 percent

Retailer's margin:
20 to 65 percent
(A regional grocery store is likely to have a 50 percent margin, and smaller specialty food stores could have 65 percent margins.)

Remember, the more handlers your cheeses move through before they get to the consumer, the higher the price will be for the consumer.

When distributors say their margin is 23 to 25 percent above FOB, it is actually higher than that. This is because the cost of freight is usually added to the FOB price before the margin is calculated. So the distributor has taken profit on the cost of freight, rather than calculating the margin from cost and adding on the cost of freight. Then the retailers add their margins (which are always high in order to cover their high cost of doing business) with the freight costs of the wholesaler compounded.

You know that your FOB price of $6 or $7 per pound is quite reasonable and that you can't be profitable below that price point, but the price at retail can more than triple once all the margins have been added together. This added margin influences how cheesemakers price their cheese for distributor sales and sell enough cheese (at the right price) to survive. Table 5.1 shows how added margins affect the final retail price of your cheese.

Now that you are familiar with margins, you can see how challenging it could be for you to sell a cheese through a distributor with an FOB price over $10. This pressure to keep the FOB price low is what makes it so difficult to be a microscale cheese producer. You may develop your cash flow models to make only 15,000 pounds of cheese a year and sell it for $35 a pound, but there are limited distribution channels for a cheese at that wholesale price. You would need to sell 288 pounds per week to meet your revenue projections.

FREIGHT

Cross-docking and *load consolidation* are terms you want to become familiar with. The goals of cross-docking and load consolidation are to reduce inventory storage, transportation costs, and transit time. They are techniques commonly used in the trucking industry.

Cross-docking is the practice of unloading goods (your

Table 5.1. Adding in margins and accounting for freight cost*

Your FOB price	Freight (per pound)	Delivered price	Retailer cost with distributor margin of 25%	Consumer price with retailer margin of 60%
$10.00	$0.75	$10.75	$14.33	$35.83
$9.00	$0.75	$9.75	$13.00	$32.50
$8.00	$0.75	$8.75	$11.67	$29.17
$7.00	$0.75	$7.75	$10.33	$25.83
$6.00	$0.75	$6.75	$9.00	$22.50
$5.00	$0.75	$5.75	$7.67	$19.17
$4.00	$0.75	$4.75	$6.33	$15.83
$3.00	$0.75	$3.75	$5.00	$12.50

Note: *Price to retailer = delivered price ÷ (1.00 − .25); price to consumer = price retailer paid ÷ (1.00 − 0.60).

cheeses) from an incoming truck and loading them into (multiple) outbound trucks with little or no storage time in between. The goal of cross-docking is to decrease product time in transit (something else that will cost you money). This is used in the cheese industry to sort cheeses intended for different destinations (e.g., breaking up a pallet of cheeses from one producer and/or combining cheeses from different origins into pallet loads).

Load consolidation works by putting together goods of different origins that are bound for the same destination. If the load wasn't consolidated, the shipment of a small cheese order could be delayed until several pallets or an entire truck was ready to go. Again, this would increase your product transit time and slow your cash flow.

For example, ABC Cheese has one pallet of 2-kilogram cheese wheels that will eventually be sold in specialty grocers in Southern California, the Bay Area, and Seattle. "Distributor A" serves the stores in Southern California and the Bay Area. "Distributor B" serves Seattle. ABC Cheese delivers its pallet to Distributor A in the Bay Area, where the pallet is then broken up into the three retail orders. The order for Southern California is loaded onto a new pallet with other southbound cheeses via Distributor A; the Bay Area cheeses are put on

Distributor A's local truck; and the Seattle order is put on a truck from Distributor B, who has an agreement to do cross-docking and load consolidation for 12 cheesemakers at the warehouse of Distributor A in the Bay Area. Without cross-docking and load consolidation, ABC Cheese might not be sold in Seattle because of the small volume of retail orders.

The cheese producer can pay a small per-pound fee for the service. Otherwise, he or she would be shipping by air or ground services, which are cost prohibitive unless this cost is passed directly to the customer (mail or online ordering).

One of the most effective ways to achieve quantities required for pallets is through consolidation of orders with other cheesemakers (into full pallets) at a central point. One possibility for consolidation is to work with an existing consolidation program, and the second is to use a distributor who already has access to good trucking rates. The first would not add cost to the shipment, but it would require negotiating a good trucking price. The second would guarantee a good trucking price, but it would mean the addition of some sort of markup on the distributor company's part to cover its expenses and ensure a profit. Networking with other cheesemakers is critical to make this a success.

Logistics can be defined as the management of goods, services, and related information from the point of origin to the point of sale or consumption. It is more than just the physical act of distribution. Here are some important questions to consider:

- Who will transport and store supplies and finished products—and how and when?

- How will inventory levels be managed—and by whom?

- Who will collect, analyze, and share data on customer orders, billing, and payment —and how and when?

- Do I have a contingency plan?

MARKETING PLANNING FOR A SMALL CHEESE PLANT

Introduction

Most people who think about getting into cheesemaking believe they will spend most of their time making cheese. In reality, marketing your cheese is going to take at least 50 percent of your time, and maybe more! Marketing your cheese is not something that you can just do in the evenings or on weekends. It is an integral part of daily operations. If you do not want to spend a significant amount of time marketing your cheese, then you will have to hire someone to do it for you. Familiarize yourself with "the Four P's" of the marketing system: product, price, promotion, and place. These elements will be covered in detail in this chapter.

If you want to capture some of that 70 to 80 percent marketing margin mentioned above, you need to create a marketing plan for your cheese. The earlier you do this, the better. Your marketing plan will represent a significant portion of a business plan. Once you have defined the objectives of the company, a marketing plan identifies appropriate markets, summarizes the marketing environment, and develops effective marketing strategies. Refer back to Chapter Three, "Building a Business Plan," to review the development of your goals and objectives.

Just like your entire business plan, a marketing plan is not a static document; it is dynamic, continuously evolving and being refined to address the current market structure and conditions. Marketing plans are generally developed to cover a relatively short period of time. Many are simply 1-year plans that are updated every year. A marketing plan for new products may cover a 3- to 5-year planning horizon, initially developed for 1 year and then extended carefully over a 3- to 5-year span. The annual updates are critical for incorporating new developments and knowledge of the market so that the strategies can be refined to

help you achieve your business objectives.

A good marketing plan forces you to be honest about your product (its quality, attributes, etc.), and realistic about the markets that you are competing in, your customers, and the return on your investment.

Six Steps for Developing a Marketing Plan

There are six broad steps in developing a marketing plan. Each step will force you to ask and find answers to difficult questions. More detailed information is available in market-planning texts that we will reference at the end of the chapter. In step five, we will discuss some tools to use for developing your marketing plan. The six steps are as follows:

1. Analyze the current situation, trends, strengths, weaknesses, opportunities, and threats

2. Understand markets and customers

3. Find your niche market

4. Determine objectives and directions

5. Develop marketing strategies and programs with specific tools (using the Four P's of product, price, promotion, and place)

6. Track progress and activities

Figure 5.3 illustrates the iterative process of marketing planning.

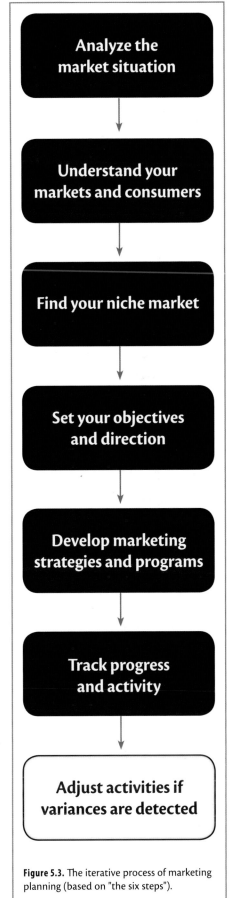

Figure 5.3. The iterative process of marketing planning (based on "the six steps").

Step 1. Analyze the current situation

The first part of an effective marketing plan consists of getting to know and understand your industry so you can see your strengths, weaknesses, opportunities, and threats (also known as SWOT).

Start with an overview of the marketplace, including demographic trends, economic conditions, technological trends, and other forces that may influence the market.

Which segments of the population eat cheese, and eat *your kind* of cheese? Go back to Chapter Three and see the suggestions about where to do market research. Don't forget to look on the Web for studies that analyze specific cheese markets.

Trend analysis

Demographic trends. Demographic trends are the most important trends to analyze. Consumer demographics look at market dynamics that influence your potential market, noting factors such as population growth, immigration, and tourism. What are the ages, gender, ethnic and religious makeup, education, household size, and income of the population specific to your market?

Economic trends. Trends in income, debt, credit availability, and usage give you a good picture of how consumers and local businesses interact in the market, and they help identify both opportunities and threats in the market. This is especially important when selling a specialty product, in both the retail and foodservice sectors.

Ecological trends. The natural environment can influence businesses in two major ways. The first is the availability and sustainability of raw materials such as energy, milk supply, and other processing essentials. The second is environmental issues in general. What environmental issues affect the specialty cheese industry because of government regulations and social attitudes? Table 5.2 illustrates consumer preferences and attitudes in this context. Although consumers state definite preferences for buying locally produced food that they perceive as being healthy for them, their shopping habits are influenced by the perception of product freshness and quality, driven by taste, and shaped by availability and access. If a product is local and organic but is not perceived as fresh and of high quality, consumers won't purchase it. On the other hand, they won't necessarily drive miles out of their way for a great product if the effort and expense are too great.

Technological trends. Changes in technology can create both opportunities and threats. If you don't use e-marketing, will you be at a competitive disadvantage? Are consumers using technologies

Table 5.2. Consumer preferences

Buying practices	Percentage of participants agreeing with the statement				
	Very important	Important	Neutral	Not important	Don't think about it
Buying locally produced foods	38.2	38.2	17.6	0	5.9
Buying food direct from family-owned farms	14.7	64.7	5.9	11.8	2.9
Buying organic foods	21.2	51.5	15.2	12.1	0
Buying foods produced in a sustainable manner	31.3	34.4	28.1	3.1	3.1
Buying foods that have potential health benefits	52.9	41.2	5.9	0	0

Source: Reed and Bruhn 2003.

to purchase cheese? If so, do you plan to have a virtual store in which consumers can learn about your cheese—and purchase it? What resources are needed to support your Web site to ensure that customers receive optimal service?

Will you need to incorporate barcoding or radio-frequency identification (RFID) for inventory tracking and shipping? What is the cost of adopting the technology and what is the benefit it brings? Can you enter the market without it?

Political-legal trends. State and federal laws make up the legal and regulatory environment in which you will do business. Political developments can signal changes in the legal and regulatory environment. Do you know whether raw milk cheeses will be outlawed or regulations relaxed? Will whey disposal be subject to additional regulations? Will individual cheesemakers be required to have a license to make cheese?

Social-cultural trends. As previously mentioned, dynamic social and cultural trends create a continuously changing market into which your product(s) will be sold. Slowly changing core beliefs about subjects like the environment, animal rights, and genetic modification may create opportunities or threats for certain types of businesses.

Competitor analysis
Analyzing competitors can help

you better understand market dynamics and anticipate how rivals react to changes in the market. You need to identify current and potential future competitors. How much of your market do they occupy? What are their unique advantages in the market? Do they have any disadvantages, particularly ones that you may use as an opportunity?

SWOT analysis
Now that you have summarized your current situation, use the summary to analyze your strengths, weaknesses, opportunities, and threats (SWOT). Strengths are internal capabilities that can help the firm achieve its goals and objectives. Weaknesses are internal factors that can prevent your business from achieving its goals and objectives. Opportunities are external circumstances that the organization might be able to exploit for higher performance. Threats, or constraints, are external circumstances that can potentially hurt the organization's performance, now or in the future.

There are two environments that you need to know and understand as part of your SWOT analysis:

- Microenvironment: Composed of mostly controllable elements that have a direct effect on the organization's ability to reach its goals and objectives.

- Macroenvironment: Composed of largely uncontrollable elements outside the organization that can potentially influence its ability to reach set goals and objectives.

Figure 5.4 provides a sample SWOT analysis and space to do your own analysis.

Step 2. Understanding markets and consumers

Understanding markets
Your overall aim in this analysis is to narrow your focus to the target market. In defining a target market, you have refined your focus to people who eat cheese, who are exposed to *your* cheese (usually limited to a geographical area), who enjoy eating your particular cheese, and who have a need or desire for your particular cheese. The target market may also include all those people who already purchase your cheese on a regular basis.

As discussed previously, only about 10 percent of all cheese consumed in the United States is specialty cheese, and about half of that is artisan or farmstead cheese. So about 5 percent of the national cheese market is artisan/farmstead cheese, which has the potential to almost double its market share. Even these small bits of information help shape the definition of a target market.

Remember, markets are always changing. Is your target

Figure 5.4. Sample SWOT analysis for a specialty cheese producer and blank SWOT analysis worksheet to fill in.

Strengths	Weaknesses
new cheese	low consumer awareness
unique attributes	no brand recognition
natural, farmstead, handmade	limited shelf life
local product	premium pricing
great taste	seasonal variation
competitive price	lack of resources
good reviews	shipping problems in warm weather

Opportunities	Threats
high consumer interest and demand	consumer loyalty
growing market	financial strength of more experienced competitors
increasing interest in health	lack of retail outlets
increasing opportunities with Internet	many other competitors
USDA health recommendations	

Strengths	Weaknesses

Opportunities	Threats

market expected to grow or shrink—and by how much? How many new businesses are expected to enter or leave the market? What are the projections for sales in your market segment over the coming years? Do these projections suggest a healthy, growing market, a stagnant market, or a shrinking market?

Understanding consumers

Many successful companies attribute their success to the fact that they understand their customers. The process of meeting underlying customer needs is at the heart of a successful marketing plan. Customer needs must be carefully researched and analyzed as the basis of all marketing-strategy decisions. A superficial analysis is not enough. While the market assessment looks broadly at who buys your cheese, you need to get a close-up look at your customer to have the complete market picture.

Consumer needs

In a marketing study done by the University of California Cooperative Extension (Reed and Bruhn 2003), consumers were surveyed on their usage

There is lots of competition in the cheese case. You need to have an excellent product that will sell the customer on the first taste. *Photo:* Barbara Reed.

and attitudes regarding specialty cheeses. Respondents consumed cheese as an appetizer, as an ingredient in salad and cooked dishes, and as a snack food. They also included cheeses as a dinner course, which has been a tradition in European households and restaurants, but is just now becoming popular in the United States.

Specialty cheese consumers stated that they are not reading labels to find low-fat and low-sodium cheeses. Instead, they look for a description of the cheese flavor and characteristics, country of origin, storage, ripening and aging information, food pairings, and "sell-by" date. Fat content confirms quality and taste attributes, and participants said they specifically avoid purchasing low-fat cheeses. These consumers are aware of the consequences of excessive consumption of any food and believe the quality of a high-fat food must be very high before they will commit an allotment of fat and calories to cheese. Many of the participants mentioned that their focus is on the quality and not the quantity of food they consume.

These consumers make no effort to economize when buying artisan cheeses, although a few of them encountered "sticker shock" when they reached the checkout stand. For this market segment, purchasing decisions are driven by the occasion, and by flavor and use. Consumers often make purchasing decisions only after sampling some cheeses.

Case cards should sell your cheese in just a few words. You need to provide your distributor with key information for the card. *Photo: Barbara Reed.*

A careful consumer analysis will help you determine which segment(s) of the consumer market to target, what strategy to use, and what product mix would be most effective. What makes consumers buy? Who is doing the buying—and what, when, where, and how do they buy? Figure 5.5 provides a summary of findings from focus group research on California specialty cheese consumers and what drives their cheese selection.

Buying decisions and behavior
No analysis is complete without research into consumer buying decisions

---

Figure 5.5. Snapshot of specialty cheese market consumer research in California.

Preference and Selection

- Consumers are not afraid of "stinky" cheese.
- They enjoy a wide variety of cheeses and have no clear favorite.
- They will listen to the recommendations of store staff or friends.
- They enjoy in-store sampling and strongly prefer to taste cheeses before purchase.
- Narrative descriptions are a powerful sales tool.
- Some consumers have a romantic vision of cheese production. They associate specialty cheeses with European cheesemaking traditions.
- Excellent customer service and sampling drive sales.
- Customers don't like pressure, and they look for flavor descriptors and information on food pairings.

Health Issues
- Specialty cheese consumers are aware of the consequences of excessive consumption.
- The quality of the high-fat foods they enjoy must be very high.
- They are focused on quality and not quantity of food, and they are concerned about antibiotics and hormones in milk and the feed given to cows.
- Consumers are interested in the potential health benefits of foods.
- Perception of quality and freshness is more important than other factors.

Price Sensitivity
- Few consumers of specialty cheese are price sensitive. When buying artisan cheeses, no effort is made to economize.

and behavior. Who in the household is doing the buying? How do consumers decide what and how often to buy? What are consumers buying, and are their purchasing patterns changing?

Figures 5.6 and 5.7 present some data about specialty cheese consumers. Forty-eight percent of the cheese consumers surveyed said that they purchase specialty cheese more than 75 percent of the time. The figures show both how much consumers purchase (as a percentage of all their cheese purchases) and how frequently they purchase specialty cheese. Specialty cheese shoppers don't buy a lot of specialty cheese (in pounds) at any

one time, but most of the cheese they buy is specialty cheese. If you are successful in connecting with dedicated specialty cheese consumers, they may become part of a loyal customer base and be exposed to your product often because of their frequent trips to the specialty cheese counter.

A memorable name and concise flavor descriptors promote cheese effectively. *Photo: Barbara Reed.*

Conducting low-cost market research on your own

One California artisan cheesemaker did some of his own market research (Pedrozo, unpublished data) while giving out samples of a product in a supermarket. He asked shoppers to rank four different "shelf talker" statements from 1 to 4 (table 5.3). A score of 1 meant the statement would be most likely to influence a cheese purchase. For the first three statements, he used descriptions that contained information about the farm family, how and where the cheese was made, and its flavor. These statements would be more likely to influence cheese purchases. The last statement (which

Figure 5.6. Consumers report specialty cheese purchases as a percentage of all cheese purchases they make.

● Specialty cheese was less than 25% of cheese purchases for 11% of consumers.

○ Specialty cheese was 25 to 75% of cheese purchases for 41% of consumers.

● Specialty cheese was more than 75% of cheese purchases for 48% of consumers.

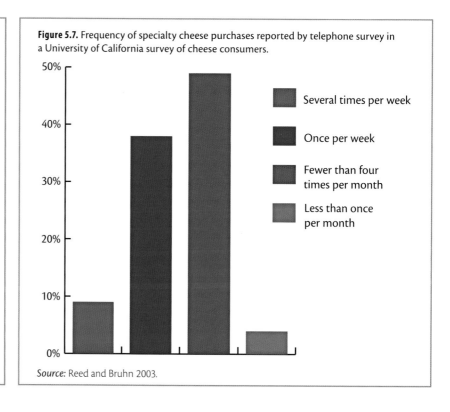

Figure 5.7. Frequency of specialty cheese purchases reported by telephone survey in a University of California survey of cheese consumers.

Several times per week

Once per week

Fewer than four times per month

Less than once per month

Source: Reed and Bruhn 2003.

Table 5.3. Shoppers' ranking of proposed case-card descriptions

Statement	Mean score
Pedrozo Peppercorn is a Gouda-style cheese, handmade in limited quantities by Tim and Jill Pedrozo. The Pedrozo family ages this cheese over 60 days. This cheese is great as an hors d'oeuvre or on top of a salad.	1.83
Pedrozo Peppercorn is a smooth, rich, Gouda-style cheese made exclusively from the milk of Jersey cows grazed on organic pastures. This semi-hard cheese has a mild flavor accented with pepper.	2.04
Pedrozo Peppercorn is a Gouda-style cheese accented with Indonesian peppercorn. It is made on a small family farm in Northern California. The milk for this cheese is produced without rBST.	2.63
Pedrozo Peppercorn is a semi-hard, Gouda-style cheese made exclusively from the milk of Jersey cows. The Pedrozo family ages this raw milk cheese over 60 days.	2.79

was evaluated less favorably) gives less descriptive information about the cheese. He did not have to pay a market research firm to do the work. A word of caution about this kind of work: don't just use a single location or small group of customers to design your entire marketing strategy or you may end up with a failed strategy. You should do your market research on as broad an audience as possible.

Business markets (distributors, foodservice, and retail grocery)
Foodservice and wholesale customers can be a significant part of your market. Business markets are influenced by individuals who buy on behalf of their company, and their decisions are influenced by different factors than those affecting the consumer.

Does a company set strategies regarding product mix and sales at the national level? Does it buy centrally? Or does it allow local stores to purchase their own inventory? Organizational considerations also include the company's size, share, growth, competitive situation, financial constraints, and timing of purchases. Doing this research will help you get on the short list of approved suppliers and open up new markets.

Step 3. Finding your niche (market)
The foundation of niche marketing is identifying segments, targets, and positioning by doing the following:

- determining the different marketplace needs and groups (segments), (e.g., specialty cheese consumers)

- targeting those needs and groups that a segment can fulfill in a unique and superior manner (targets), (e.g., aged blue cheese with a natural rind)

- positioning products, attributes, and services so that the target market recognizes the company's distinctive offering and image (positioning), (e.g., organic farmstead cheese)

No company or product can be all things to all customers, so use segmentation, targeting, and positioning to narrow the focus of the marketing plan.

Segmentation
It is not possible to service an entire market with a single product or product attribute that will satisfy everyone, nor should you try. There are pasteurized process cheese foods, spreads, modified-fat cheeses, soy and hemp "cheeses," "commodity" cheeses like Cheddar and Jack in 50-pound blocks, fresh cheeses, aged cheeses, European cheeses, flavored and herb cheeses, etc.

Cheeses are often featured with other specialty foods such as olive oils and chocolates. *Photo:* Barbara Reed.

Know your product attributes so you can identify those population segments that will purchase the attribute and that are within a market in which you have the ability to compete.

By identifying a specific market segment, you will find niche markets—smaller segments within a market that exhibit distinct needs or benefit requirements. Artisan and farmstead cheeses are niche products within the specialty cheese segment. Over time, small niches can grow into very large ones, and they can become highly successful segments for the companies that serve that segment. Even if your niche includes a large consumer segment (e.g., Whole Foods shoppers), you may decide to serve only some of these consumers by selecting a specific geographic area, close to large metropolitan markets such as the San Francisco Bay Area, Atlanta, Chicago, and New York.

Targeting

Once you have identified your relevant segments within the

overall market, evaluate each segment (consumer group) in light of your goals and objectives as well as your offerings. One key measure of the segment is current and future opportunities for sales and profits. A second measure of the segment is the potential for competitive superiority. A third measure is the fit with your resources and core competencies. A fourth measure is the extent of environmental effects (regulations, etc.) and the degree of difficulty of entering the market.

Prioritizing market coverage

After choosing segments for targeting, you need to rank them and arrange them in order of priority for entry. Your marketing plan should explain your marketing priorities. There is no particular way of doing the actual ranking. The most useful criteria are the ones you used to select target segments in the first place: current and future sales and profits, competitive superiority, fit with resources, and environmental threats, etc.

Position

Designing the company's offering and image to occupy a unique place in the minds of consumers in a target market is called positioning. When you do an excellent job of positioning, you can develop the rest of your marketing planning and differentiation from this position of strength.

"The best unexpected thing that has happened has certainly been our rate of growth, beyond all expectations."

Cheesemaker Story

Step 4. Determine Objectives and Direction

A formal way to communicate objectives and direction is through a mission statement (see Chapter Three, "Building a Business Plan").

Strategic direction

The ultimate purpose of the marketing plan is to help your cheese business achieve objectives (i.e., shorter-term performance targets) that will, in turn, bring it closer to achieving its goals (i.e., longer-term performance targets tied to the mission).

You put a lot of work into your mission and direction in Chapters Two and Three. Figure 5.8 shows strategic direction options. Although there may be short plateaus during your growth, when you are able to stop for a moment and catch your breath, most of the time you should be growing—and continuing to focus on market penetration, development, product development, and diversification. These strategies are outlined as follows:

Organic cheeses are a specialty category within artisan cheeses. *Photo: Barbara Reed.*

- **Market penetration.** Create a growth strategy in which the company sells more of its existing products to customers in existing markets or segments.

- **Market development.** Identify and tap new segments or markets for existing products.

- **Product development.** Sell new products to customers in existing markets or segments.

- **Diversification.** Offer new products to new markets through internal product development or by starting (or acquiring) a business for diversification purposes.

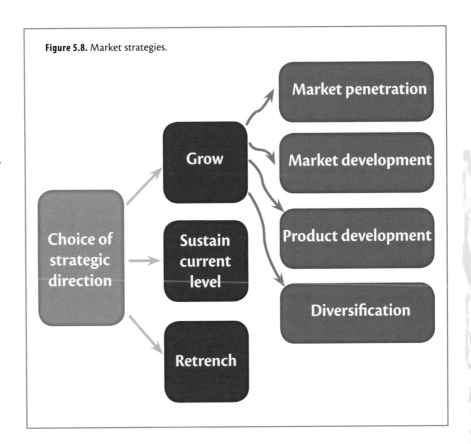

Figure 5.8. Market strategies.

Setting objectives for your marketing plan

Once you have determined the strategic direction, you can review the SWOT analysis to identify potential pitfalls and sources of strength for achieving marketing plan objectives. These objectives should include expected financial results.

Objectives will be effective for guiding marketing progress only if they are quantitative measures such as sales figures, customer counts, or satisfaction surveys for a given time period.

Objectives should be

- realistic but challenging

- consistent with the mission and overall organizational goals

- consistent with internal resources and core competencies

- appropriate in the light of external environmental opportunities and threats

Figure 5.9 illustrates the linking of strategy, goals and objectives, tactics, and programs.

Step 5. Marketing strategies and programs

Marketing strategies and programs are the heart of the marketing plan. They are based on the market analysis, the objectives of the marketing plan, and the segmentation, targeting, and positioning of the product.

Your strategy must include specific actions and tactics. For example, setting up timetables to show the starting and completion dates for each activity, assigning responsi-

"Walk before you run. Perfect one or two cheeses before you expand your product line."

Cheesemaker Story

bilities for individual tasks, determining actual prices at which the product will be sold, devising promotion actions, etc. You have five main tools that you can use to create your marketing plan, and these are "the Four P's" plus customer service. The "Four P's" of marketing are: product, place, price, and promotion.

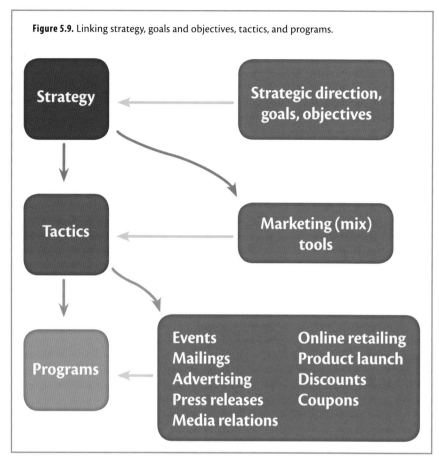

Figure 5.9. Linking strategy, goals and objectives, tactics, and programs.

Strategy ← Strategic direction, goals, objectives

Tactics ← Marketing (mix) tools

Programs ← Events, Mailings, Advertising, Press releases, Media relations, Online retailing, Product launch, Discounts, Coupons

example, full refunds or coupons for free product if the customer has a problem with your product). Credit terms and delivery are important to your wholesale or online customers.

Product classifications

Specialty cheese is definitely a perishable and consumable food that will have limited points of purchase (you won't find it in a 7-11), but it may be an impulse purchase with little or no price comparison.

Product mix strategy

You should start out with a simple and manageable product mix. The dimensions of a full product mix include

» width (number of different product lines—for example, aged and fresh cheeses)

» length (number of items sold in a product line—for example, four types of fresh cheeses)

» depth (variations of items sold in the product line—for example, six different flavorings with herbs, or garlic in each fresh cheese line)

The above three factors give you a matrix of options for market expansion.

Figure 5.10 is a worksheet that lets you develop your own marketing tools (using the Four P's).

Product

Your main marketing tool is your cheese. But you also need to think about the entire offering: the cheese itself and the value and service you include with it. This may include packaging and information about your cheese. And are you willing to replace, free of charge, any cheese that doesn't meet a customer's expectations?

We usually think of a product as a physical item. But a product can be anything that we are selling, including terroir (the local growing conditions that help create your unique product), organic production practices (and product), a cheese "handmade by traditional methods," and your family farm identity.

Product components

Most products have both tangible and intangible characteristics, e.g., color, shape, price, style, and quality. A product usually has a brand name that is used to identify it. Packaging may be important, depending on the market segment. Guarantees are often important (for

Packaging small-weight wheels can be a little less labor intensive than having to cut cheeses to exact weight and thent wrap them. *Photo:* Steve Quirt.

Figure 5.10. Using your marketing tools.

The Four P's

Product: the definition of your product

Price: the cost, price, and placement of your products

Promotion: how you inform your customers of your products and how you sell your products

Place: where you sell your products

Write examples here that are specific to your cheese business:

Product	Price	Promotion	Place

Brand strategy

Brand strategy is concerned with deciding which products should be branded, and whether they should be sold under your own label or under labels controlled by other firms (co-packing).

The American Marketing Association (AMA) defines a brand as "a name, term, sign, symbol, or design, or a combination of them, intended to identify the goods or services of one seller or group of sellers and to differentiate them from those of competitors."

Brand is the focal point of your product strategy. Creating, sustaining, protecting, and enhancing brands are distinctive skills of marketing professionals. The seller of a product or service is granted exclusive rights to the use of the brand name in perpetuity (forever) under trademark law. A brand can convey many meanings, including attributes, benefits, values, culture, personality, and type of user. Ultimately, branding lies in the perception of customers, so it is critical for a company to research the position that its product(s) or service(s) occupy in the minds of customers. You need to make careful decisions regarding the brand's name, logo, tag line, type style and colors, or symbol, which bring about recognition (brand identity) from current and prospective customers. Get help from an experienced marketing professional to assist you with this task.

This process is more than advertising—it should be backed up by delivery of a consistent product or service that brings real value to your customer. When customers have a completely positive experience with a company and its product, it's called brand bonding. Successful branding implies the product's benefits and qualities. Furthermore, it is easy to recognize and remember, and it is free from negative connotations.

Product quality

Your ability to compete in the specialty cheese market will be directly linked to the quality of your product. No one can afford to buy poor-quality specialty cheese.

Features and benefits

Features are specific attributes that hold your product in its intended niche (for example, as cheese that has an ash layer and is wrapped in a fig leaf). Benefits are "need-satisfaction outcomes" that customers desire from the product. Examples might be a goat cheese product that is easier to digest or a product whose package size or type is convenient for the consumer. Table 5.4 gives a snapshot of cheese aging and packaging from a subset of California cheesemakers. It tells us about the form of the product they sell.

Pricing strategy

Finding the right price

Many people think that all you do is estimate your costs, add a margin for overhead and profit, and voilà, you have your selling pricing! But you need to recognize that the amount you can sell varies according to the

Table 5.4. Producer survey on cheese aging and packaging*

Production steps and details		Quantity of cheese produced (lbs)	Percent of cheese produced (%)
Aging	not aged	396,280	34.5
	2–5 mo	312,300	27.2
	6–12 mo	341,720	29.7
	23–24 mo	93,800	8.2
	≥ 2 yr	4,700	0.4
Packaging	cut and wrap at plant	392,928	34.2
	leave as whole wheel	755,872	65.8

Note: *A survey of some California cheesemakers illustrates that the majority of cheese made by artisan producers is sold at a few months of age and is not elaborately packaged. Moving inventory helps cash flow, and selling whole wheels eliminates the need to own packaging equipment.

price you set. And costs often change with volume, so profits depend on both price and costs. If customers compare the perceived quality of your goods and services against price (value), how do you price? Some customers want the highest quality for the lowest price, while others are willing to pay for high quality. Therefore, your price must be in line with your customer base, and it must be reasonable when compared with your competitors.

You must also be careful to consider how you price all of the items you are selling in your line. The price of one product you sell can also affect the sales of another product you sell.

Some consumers may even be suspicious of a product that is priced below a competitor. In an unpublished University

"Make unique cheeses so you don't have to enter the artisan cheese market by having the lowest-priced product to get placement. You will need a certain amount of sales to cover your costs; be sure you have the space to make the cheese you need to sell. Decide ahead of time what you need to sell, based on what your costs will be."

Cheesemaker Story

of California survey about organic cheeses, some respondents said they would be suspicious of an organic cheese selling at a lower price than regular cheese or at a discounted price compared with other organic cheeses.

Determining pricing goals

Your first task is to determine your overall pricing goals. If, for example, you are the first into the market with a unique cheese, then you are

in a very good position to follow a premium pricing strategy. Why? Because you have a product that no one else makes. This is a monopoly. In contrast, in a highly competitive market the primary goal might be to maximize market share. In the short run, the only way to do that is to sacrifice revenues and/or profits, and maybe even take a loss. For most cheesemakers, sacrificing revenue and cutting prices won't be an appropriate strategy for a small start-up company. See Chapter Six on the risk of growing too fast.

» *Profit maximization*: Most firms try to maximize profits, and therefore it is the most common pricing goal.

» *Revenue maximization*: Often firms that are looking for sales growth or trying to increase market share use an alternative to profit maximization called revenue maximization. Often, you can sell your product at a lower price—but in greater quantity—when you maximize revenues; but also be aware that you are sacrificing profits when you do this. It is not recommended for a small-volume producer.

» *Market share maximization:* This objective trades market share for profits and revenue. In this case, you would drop the selling price to as low as possible, and try to drive competitors out of the market. Again, this isn't a smart strategy for a specialty product. You aren't very special if your product is cheap and available everywhere.

» *Quality leadership:* Often buyers are willing to pay more for higher-quality goods. In this case, if you truly have a superior (and unique) product, then you should push this line and make quality your overall goal. But you should also be aware that quality is often a perception in the consumer's mind, and if a competitor also can show that their product is just as good at a lower price, you will lose market share and revenue, and eventually you may be priced out of the market.

"One of our biggest challenges is the ever-escalating high costs, the labor-intensiveness to produce the product, and then trying to keep our costs down so that our cheese is competitively priced in the marketplace."

Cheesemaker Story

Table 5.5 provides some sample pricing objectives with both financial and marketing implications. You may set prices to return a specific profit margin, or to make a return on your investment, or to reach a percentage of market share or number of customers.

Measuring demand

There is a relationship between price and demand. That is, as the price moves up, the quantity demanded moves down. Almost every purchasing decision you make in your everyday life follows this general rule. You will learn the price points where demand decreases for your cheeses (price elasticity of demand) or have a good idea what your demand curve looks like. You only need a few observations to make a good stab at your demand schedule.

Estimating costs

Costs have two basic components: fixed and variable. Fixed costs are the costs of all overhead, plant, equipment, salaries (if labor is not variable), and all of the costs that are sunk in setting up the business but before anything is produced. Variable costs are those associated with actually producing a good, such as the materials, variable labor, commissions, etc. Total cost is the sum of fixed costs plus variable costs.

Excerpt from marketing research interviews with retailers:

Observations: "The breakpoints in price, by which is meant the price at which you can charge without hurting volume, are retails of $9.99, $11.99, $14.99, and then $19.99. Depending on the perceived value of the cheese, retailers will sell the most cheese at $9.99, but lower prices will not improve sales. They will sell slightly less volume at $11.99, and slightly less again at $14.99. The drop-off from $14.99 to $19.99 is considerable—and just like on the West Coast, cheeses sold at above $19.99, at retail, slow down considerably . . . [W]hen cheeses arrive on the retail shelf at $14.99 . . . this is where you maximize profit and sales, which leaves more profit dollars in the bank."

Table 5.5. Sample pricing objectives

Type of objective	Sample pricing objectives
Financial	To support profitability: set prices to achieve a gross profit margin of 40% on this year's sales.
	To support return on investment: set prices to achieve a return on investment (ROI) of 18% for the full year.
	To cover costs: set prices to break even on sales within 2 months.
Marketing	To support higher market share: set prices to achieve market share of 7% within 6 months.
	To support higher sales: set prices to achieve a sales increase of 12% over last year.
	To support customer acquisition: set prices to attract 1,500 new customers in first 6 months of year.

Pricing method

» *Markup pricing (cost plus margin)*: A specific profit margin has been calculated into the price. The savvy business person has considered price elasticity to be sure the markup will not decrease demand.

» *Break-even pricing*: At a given number of units sold (break-even volume), your unit price will cover your costs. You should know your break-even price and volume at all times, especially before you consider selling to higher-volume customers such as Costco.

» *Target return pricing:* ROI (return on investment) is pricing calculated to cover all costs and achieve a target return. It is less specific than cost plus margin.

» *Variable cost pricing:* This uses variable costs to calculate minimum pricing, without accounting for overhead (fixed costs).

Discounts and allowances

You should factor into your pricing the periodic discounts and allowances that are expected by distributors and retailers. In the specialty cheese market, you should be able to avoid "price-only" competition by highlighting the value and attributes that are exclusive to your product.

Place

Place strategy lays out the distribution methods and partners you will use to reach customers in the target markets you have selected. Some cheesemakers use direct sales (farmers' markets, on-farm stores), while others may use brokers or distributors. Others sell to specialized outlets, retail stores, foodservice, or restaurants. The specifics of place strategy were discussed in detail at the beginning of this chapter.

When you encounter a retailer who wants to carry your cheeses, carefully consider the retailer's operation and clientele to ensure that this market

outlet is consistent with your positioning. For example, if you produce highly perishable fresh Mozzarella cheeses and the retailer carries a wide variety of aged Cheddars, the outlet may not be a good match for you.

Promotion

Promotion covers all the tools used to communicate with the target market, which include deciding on advertising objectives, developing an advertising budget, developing campaign issues and themes, choosing appropriate media, and monitoring results. You need to think carefully about how you are going to promote your cheese. There are some simple but labor-intensive ways of promoting your cheese, such as selling at farmers' markets, providing samples at food and wine festivals, or doing in-store sampling and demonstrations. And then there are some more costly ways of doing it, like hiring a broker or creating advertising mailers. Whatever methods you use, it is essential to carefully manage and monitor the overall content and impact of your promotion strategies to ensure that your messages are getting to the right markets and the right people.

Advertising objectives

New products require quite different advertising strategies compared with those that are in a growth phase, in the maturing phase, or in their declining phase. For example, when introducing a new cheese, you may want to have a big promotion and provide incentives to wholesalers to give it the exposure you want. A product being phased out may have deep discounts to clear it from your inventory.

Sales promotion

Small and medium producers will not use mass media advertising to any extent. There are, of course, a number of ways of promoting your product other than paid advertising. These can be thought of in the following categories:

» *Product promotion: samples, bonus product, premiums.* You will use this method the most. Sampling is what drives sales in the specialty cheese marketplace. Providing cheese to food events, brokers, and distributors is part of this strategy.

» *Price promotion: coupons, discounts, refunds, credit, etc.* This method will be used with distributors and retailers. You may provide discounts on large-volume orders, a special price on one item, or credits if sales exceed a certain volume over time.

» *Distribution promotion: trade shows, point-of-purchase materials.* If you plan to sell beyond farmers' markets and direct sales at your store, you will want to participate in trade shows like the Fancy Food Show in San Francisco or on the East Coast, and you will need to develop sales materials for your distributors to share with retail outlets.

» *Communication promotion: personal selling (demos, trade shows, exhibitions, competitions).* This could be done, for example, through the Fancy Food Show, American Cheese Society, and World Cheese Awards.

» *Publicity: special events, press bulletins, press conferences, tours, etc.* When you introduce a new cheese or expand into a new market area, send out a press release to trade publications such as *Gourmet News* or *Cheese Market News*, the consumer food media, and to area newspapers. You may be invited to appear on a local TV station and have the station personalities sample your cheese. It is a good idea to develop your own media list of key writers and reporters. Developing relationships with specific reporters and writers is far more effective than just sending out press releases to a general list of publications and news outlets.

Table 5.6 provides a snapshot of promotional tools, and they are listed in the order you would be most likely to use them.

Table 5.6. Major promotion tools

Tool	Use	Examples
personal selling (direct sales)	reach customers one-to-one to make sales, strengthen relationships	farmers' markets, sales appointments, sales meetings, presentations
public relations	build positive image, strengthen ties with stakeholders	event sponsorship, news releases and briefings, speeches, public appearances
sales promotion	stimulate immediate purchase, reward repeat purchases, motivate sales personnel	samples, coupons, demonstrations, trade shows, trade incentives
direct marketing	reach targeted audiences, encourage direct response	e-mail marketing services, printed and online catalogs, telemarketing, direct-mail letters and brochures
advertising	get messages to large audience efficiently	TV and radio commercials, Internet banner ads, magazine and newspaper ads, product and company brochures, trade advertising, CD- and video-based ads

Free social media sites

Facebook (www.Facebook.com) is a social networking site that connects friends and families. The Web site currently has more than 700 million active users worldwide. MySpace is another popular social networking site.

Twitter (www.twitter.com) is a free information networking and micro-blogging service that enables its users to send and read messages known as "tweets." Tweets are text-based posts of 140 characters displayed on the author's profile page and delivered to the author's subscribers, who are known as "followers."

YouTube (www.youtube.com) is the place to upload videos about your events, testimonials from customers, a virtual tour of what visitors will see at your creamery, and much more. You can then post a link to the video on your Web site, blog, or with Twitter.

Digg (www.digg.com) is a social news Web site for people to discover and share content from anywhere on the Internet by submitting links and stories, and then voting and commenting on those links and stories.

Stumble Upon (www.stumbleupon.com) is an Internet community that allows users to discover and rate Web pages, photos, and videos. It is a personalized recommendation engine that could be very useful for your operation.

Delicious (www.delicious.com) is a social bookmarking Web service for storing, sharing, and discovering Web bookmarks.

Reddit (www.reddit.com) is a source for what's new and popular online. Users can vote on links that they like or dislike, help decide what's popular, or submit their own links.

TripAdvisor Media Network (www.tripadvisor.com) is the largest travel community in the world, with 7 million registered members and 15 million reviews and opinions from travelers.

Using social media

Social media is quickly becoming the most effective marketing tool for small business. According to statistics from the California Travel and Tourism Commission, we know that

» 86 percent of Americans travel with their cell phones, which they use to call ahead to see what's blooming on the farm today or to book an experience

» 70 percent of 15- to 30-year-olds use social networks such as Facebook to learn about and share with friends. This usage is growing with older travelers as well.

» 75 percent of Web users trust online reviews more than other written sources.

Being visible is paramount. Posted customer review comments and ratings are important, and most of all, the visual appearance of your Web presence is crucial, whether it's on your Web site, a Facebook page, your blog, or a Twitter account.

While we know that the Internet is the Number One source of travel planning and purchasing, it's the consumer who is becoming the medium or gateway to your creamery or dairy via social media and networking sites. The Web site TripAdvisor, which is made up of travelers' reviews, is used by one out of four travelers; blogs about your site are also popular sources. Randall Travel Marketing predicts that this consumer-to-consumer style of travel information sharing will be one of the largest trends to impact the travel and tourism industry in the near future. Simply put, the consumer is now in control of artisan-food tourism marketing.

If you think Twitter is a type of bird, a blog is a low spot on your farm, and a Facebook page is something you see at the post office, then you need to educate yourself. Plan to attend a regional or national cheese guild workshop about these topics. The North American Farmers' Direct Marketing Association and the American Cheese Society are great resources, as are your local cheese guild and tourism bureau.

"The social media revolution is radically changing how direct-marketing farmers communicate with their customers," said Michael Straus, founder of Straus Communications and former vice president of marketing at Straus Family Creamery. "However, it's important to select the right tools for your marketing strategy; otherwise, you could risk a lifetime in tweeting with insignificant results."

"The media has had a large part to do with our sales. The best advertising definitely comes free with a well-known writer on your side!"

Cheesemaker Story

Using Facebook to advertise

Small, niche farm products can be highlighted on Facebook. Use the shop function on Facebook to create a fanstore if you plan to ship or sell products by mail.

"[By] using Facebook we are interacting with our customers/fans in a much more direct and immediate manner," comments Michael Zilber, store manager for Cowgirl Creamery. "And from a purely commercial standpoint, we are able to keep them informed on our latest products, specials, and events. But more importantly, we can use it to further our company philosophy and outreach, which helps extend the brand in general. By posting about a variety of subjects related to other cheesemakers, artisan cheese in general, and sustainable agriculture, we are furthering content that supports Cowgirl and the issues we think are important to our business."

Use Facebook's reviews wall to post visitor comments. Gather your visitors' e-mail address when they come and ask them to sign on as a fan. Blogging from your Web site or Facebook is another great way to keep your fans and customers connected to you.

So now you're wondering, which one do I set up? The answer is, as many as you can keep up to date! Your Web site and Facebook should sync seamlessly, picking up friends and fans from Facebook and customers via your Web site.

The cheesemaker should have a Web site linked to a Facebook page and a Twitter page. New forms of social media will emerge and become increasingly important to start-up cheesemakers. Be sure to integrate it into your business model. You can use your monitoring strategy to assess how well social media works for you.

Review the market customer analyses and plan for additional marketing research, if needed, to learn more about the target audience.

Monitoring the effectiveness of advertising

It is almost as important to monitor the effectiveness of the advertising and promotion campaign as it is to advertise and promote in the first place. The bottom line is: How will you know how effective your program is if you do not monitor it?

Customer service

No marketing strategy is complete without a strategy for customer service and a strategy for internal marketing. Almost all businesses incorporate some form of customer service with their product offering. The label of nearly every food product in the supermarket contains an invitation to call toll-free with questions or comments, and this is only one form of customer service. Other forms of customer service include rebates, guarantees of rapid and personal service, quality follow-up, guaranteed delivery, etc. In many cases, customer service can make or break an offering. Clearly, customer service strategy is a critical part of any marketing plan. But it also involves extra cost.

Customer service at point of purchase. Customers often need help in obtaining information about your cheese, its use, its flavor and texture, and matching the right cheese to other foods. Shelf talkers and other point-of-sale materials help, as does training sales staff in stores where your cheese is sold.

Customer service after sale. After the purchase has been completed, customers should understand how to get help if they encounter any problems with your product. See the section "Food-Safety Strategy No. 9" in Chapter Six.

Internal marketing strategy

Employees and other people who directly handle the product are goodwill ambassadors for your cheese and your company. Make sure that every person in the marketing chain (this includes plant workers and management) does their best work every day to maintain the integrity of your cheese and your company.

Figure 5.11. Dealing with rapid business growth.

Business growth can be very rapid. If you haven't planned for it, you can find yourself in a very tight spot. These issues limited the ability of some California cheesemakers to expand. They could not increase production to meet market demand unless they solved these constraints.

Barriers to Expansion
- lack of storage space and cost of space
- lack of processing space
- inadequate cash flow and excess financial risk
- lack of milk supply
- lack of expertise in cheese aging
- distribution problems (including lack of cross-docking and drop points)
- cost and availability of labor (including housing)

There are some downsides to successful marketing, and rapid business expansion has its own challenges. Figure 5.11 summarizes some of the challenges of expansion faced by California cheesemakers.

Step 6. Track progress and activities

A marketing plan must establish realistic forecasts, budgets, and schedules in order to measure progress toward objectives and pinpoint problems to be corrected. The following worksheets (figures 5.12 to 5.16) provide you with the framework to do some market assessments.

Figure 5.12

Our Features and Benefits (Examples)

Features
- available at our farm store
- aged with a natural rind

Customer benefits
- our cheese is the freshest
- best selection and price
- meet the cows
- no waxes or other coatings to interfere with the natural development of cheese flavor and texture

Our products' features and benefits

(from Chapters Two and Three and figure 5.10 of Four P's)

Features _____

_____

_____

_____

_____

_____

_____

_____

_____

_____

_____

_____

_____

_____

_____

_____

_____

_____

_____

_____

_____

Benefits _____

_____

_____

_____

_____

_____

_____

_____

_____

_____

_____

_____

_____

_____

_____

_____

_____

_____

_____

_____

_____

Figure 5.13

Identifying Our Competitors

Competition Name and Address	*How They Compete*
1. _____	1. _____
_____	_____
_____	_____
_____	_____
2. _____	2. _____
_____	_____
_____	_____
3. _____	3. _____
_____	_____
_____	_____
4. _____	4. _____
_____	_____
_____	_____
5. _____	5. _____
_____	_____
_____	_____
6. _____	6. _____
_____	_____
_____	_____
7. _____	7. _____
_____	_____
_____	_____
8. _____	8. _____
_____	_____
_____	_____

Figure 5.14

Comparing Our Enterprise with Others

Rate yourself and your competitors on a scale of 1 (poor) to 5 (excellent)

Features/Attributes	Our enterprise rating	Competitor's enterprise rating (numbers keyed to fig. 5.13)							
		1	2	3	4	5	6	7	8
Overall quality									
Flavor									
Presentation									
Consistency of product batches									
Variety of offerings									
Retail price point(s)									
Availability									
Marketing									
POS materials									
Customer service									
Brand/image									
Physical plant (equipment, technology, quality)									

Figure 5.15

Changes for a Competitive Advantage

_____

_____

_____

_____

_____

_____

_____

_____

_____

_____

_____

_____

Figure 5.16

What Makes Our Cheese Unique?

_____

_____

_____

_____

_____

_____

_____

_____

_____

POINTS TO REMEMBER

➻ A marketing strategy is critical to the success of every artisan cheese enterprise.

➻ A marketing plan is what you do to get your cheese out the door in the most efficient, cost-effective method, considering your unique niche, distribution channels, demographics, and your ultimate profit margin goal.

➻ A marketing strategy has several key components: the market, your product's unique features and story, the message, promotion and advertising, and your competitive advantage.

➻ An easy-to-use Web site is an absolute necessity.

➻ Good relations with neighbors, local businesses, cheese distributors, food writers, and others are essential to the success of any business enterprise.

REFERENCE

Reed, B., and C. M. Bruhn. 2003. Sampling and farm stories prompt consumers to buy specialty cheeses. California Agriculture 57(3): 76–80.

Chapter Five adapted in part from University of California Cooperative Extension Farmstead Cheesemaking Workshops, 2003–2005. Glenn County Cooperative Extension, Orland, CA.

The sections "Retail" and "Understanding Consumers" adapted in part from Reed, B. and C. M. Bruhn. 2003. Sampling and Farm Stories Prompt Consumers to Buy Specialty Cheeses. California Agriculture 57(3): 76–80.

The sections "Distribution" and "Freight" adapted in part from Reed, B., J. Graves, and E. Carlson. 2004. Feasibility Study of Forming a California Cheese Aging Cooperative. Final Report to the USDA Rural Cooperative Development Program.

The sections "Distribution," "Margins," Freight," and "Measuring Demand" adapted in part from Reed, B. and D. Strongin. 2005. Specialty Cheese Distribution Cooperative - Phase 2 of Cheese Aging Cooperative Feasibility Study. Final Report to the USDA Rural Cooperative Development Program.

The sections "Social Media" and "Points to Remember" adapted from George, H. and E. Rilla. 2011. Agritourism and Nature Tourism in California. 2nd ed. Oakland: University of California Division of Agriculture and Natural Resources, Publication 3484.

CHAPTER *Six*

Risk Management

∽

Barbara Reed

Chapter Goals 80

Introduction to Food-Safety Strategies. 80

Twelve Food-Safety Strategies 83

    1. Establish a training program 83

    2. Implement good manufacturing practices (GMPs) 83

    3. Follow GMPs every day. 83

    4. Develop and follow a master sanitation plan 85

    5. Conduct regular environmental swabbing 95

    6. Develop and follow a pest management plan 98

    7. Develop and follow a chemical control plan 100

    8. Establish an allergens protocol 101

    9. Establish a customer and consumer complaints protocol 103

    10. Develop a food recall plan. 105

    11. Establish process control protocols 112

    12. Obtain product liability insurance 114

Management and Marketplace Risk. 116

Points to Remember. 118

References 118

INTRODUCTION TO FOOD-SAFETY STRATEGIES

Risk management is an integral part of developing and operating a successful business. If you didn't anticipate problems, you would always be managing from a reactive position. Because you will be selling food to the public, your risk-management plan begins by developing strategies that will decrease the risk of foodborne illness and product contamination. The desired outcome of having food-safety strategies is to produce clean, safe, and wholesome food of high quality. The owner of a small business is actually in a better position to have strict process controls than a large manufacturer. You are more intimately associated with every aspect of production and will have a closer working relationship and more direct line of communication with your employees. Basic lab equipment and supplies are affordable even for small businesses, and they will help improve product quality.

Once you have mapped strategies for food safety, you also need to consider management and marketplace risk. Product and customer diversity is one way to manage marketplace risk. We will discuss this in more detail at the end of this chapter.

In a 2004 study commissioned by the FDA, independent experts identified "deficient employee training," "contamination of raw materials," "poor plant and equipment sanitation," and "poor plant design and construction" as the top four food-safety problems faced by food manufacturers today. According to a committee convened to review food-manufacturing practices, the most frequently mentioned preventive controls that apply to food-safety problems include the following:

- **Training**
  Ongoing and targeted training on issues ranging from allergen control, cleaning and sanitation procedures, incoming ingredient receipt protocol, and monitoring for employees and management, as well as suppliers.

- **Audits**
  Periodic audits and inspections of facility and raw material suppliers, either in-house or by third-party firms.

- **Documentation**
  Documentation of training activities, raw material handling policies and activities, cleaning and sanitation, receiving records, and use of sign-off logs.

- **Validation/Evaluation**
  Evaluation of training effectiveness, establishment of accountability, and validation of cleaning through testing (e.g., swabs, organoleptic evaluations, and bioluminescence tests).

In this chapter, we will discuss 11 food-safety strategies that will help you build these preventive controls into your business. Get started on the right track by making them part of your routine food-manufacturing process. Although this publication is not about cheesemaking practices, these preventive controls must be included to effectively cover risk management.

Table 6.1 shows that the most common manufacturing problems contributing to foodborne illness can be corrected with effective management tools such as the four listed above. If you can't afford third-party audits, you can monitor your own processes, in effect creating an audit of your own systems.

Table 6.1. The top 10 problems that contribute to foodborne illness and their solutions

Problem	Solution
contamination during processing	separation of production lines, use of physical detachments and lockouts, use of staging areas, routine maintenance of manufacturing equipment, and properly conducted, unbiased, third-party audit of GMPs
contamination of raw materials	supplier audits, raw material testing and verification, supplier training, preprocessing treatments (i.e., pasteurization, irradiation, washing, culling, etc.), documentation from suppliers certifying safety of materials, and properly conducted, unbiased, third-party audit of GMPs
deficient employee training	provision of training specific to the employees' duties; bilingual training; provision of learning aids such as newsletters, posters, and videos; seminars and employee reviews; evaluation of the effectiveness of training; training refresher courses; in-house training (versus consultants); and properly conducted, unbiased, third-party audit of GMPs
difficult-to-clean equipment	environmental sampling, cleaning of areas prone to niches, SSOPs, disassembling equipment to clean, addition of a kill-step at the end of processing (i.e., pasteurization, irradiation, etc.), and properly conducted, unbiased, third-party audit of GMPs
incorrect labeling or packaging	institution of label review policies, removal of old label and packaging inventories from the manufacturing site, verification of labels by scanning barcodes, label audits, training, and properly conducted, unbiased, third-party audit of GMPs
no preventive maintenance	preventive maintenance plan, documentation of repairs and servicing, and properly conducted, unbiased, third-party audit of GMPs
poor employee hygiene	use of sensor-equipped towel dispensers, keypad controls for hand washing, automated hand-washing stations, and properly conducted, unbiased, third-party audit of GMPs
poor plant and equipment sanitation	keypad controls that keep track of hand washing, sensor-equipped hand towels, pay incentives, beeping dispenser to ensure adequate hand-washing time, documentation of hygiene activities (i.e., logs), SSOPs, and properly conducted, unbiased, third-party audit of GMPs
poor plant design and construction	properly conducted, unbiased, third-party audit of GMPs
postprocess contamination at manufacturing plant	environmental sampling, inclusion of a kill-step at the end of processing (i.e., pasteurization, irradiation, etc.), use of preservatives, SSOPs, and properly conducted, unbiased, third-party audit of GMPs

Source: Adapted from a Delphi Study, FDA 2004.

Summary of Food-Safety Strategies

Food-safety strategy no. 1:
Establish a training program for yourself and your employees covering good manufacturing practices (GMPs) and safe food handling. Then establish a continuing education program to keep everyone up to date.

Food-safety strategy no. 2:
Design and set up your plant so that you can implement good manufacturing practices and so they will be easy to accomplish.

Food-safety strategy no. 3:
Follow good manufacturing practices every day in your plant.

Food-safety strategy no. 4:
Develop and follow a master sanitation plan, including

standard sanitary operating procedures (SSOPs).

Food-safety strategy no. 5:
Conduct regular environmental swabbing in your plant to verify the effectiveness of your sanitation practices.

Food-safety strategy no. 6:
Develop and follow a pest-management plan. If necessary, hire an outside firm to implement it for you.

Food-safety strategy no. 7:
Develop and follow a chemical control plan. An important element of food safety is preventing contamination of food through the appropriate use, storage, and labeling of all chemicals used in the processing facility.

Food-safety strategy no. 8:
Establish a protocol for handling ingredients considered to be allergens. This will include your product-labeling practices.

Food-safety strategy no. 9:
Establish a protocol for dealing with customer complaints. If there is any kind of problem with your product, you want to know about it as soon as possible.

Food-safety strategy no. 10:
Develop a food recall plan and practice going through a mock recall to be sure the plan works the way you expect it to.

Food-safety strategy no. 11:
Implement process controls. Sample every batch of cheese at the appropriate time to be sure you are reaching the pH, salt, and moisture content described in your make process. These three parameters are critical for inhibiting undesirable microbial growth and reducing the risk of foodborne pathogens surviving in your cheese.

Food-safety strategy no. 12:
Obtain product liability insur-

All your employees, even those who milk your animals, should understand the importance of producing clean, safe food. *Photo: Steve Quirt.*

Personnel are attired with proper footwear as well as disposable aprons and gloves. *Photo: Steve Quirt.*

ance. Protect against financial loss arising out of the legal liability resulting from the use and consumption of your product by purchasing liability insurance. In addition to liability insurance, manage risk by selling in diverse markets, planning for growth, and developing contingency plans.

TWELVE FOOD-SAFETY STRATEGIES

Food-Safety Strategy No. 1: Establish a Training Program

Establish a training program for yourself and your employees covering good manufacturing practices and safe food handling. In addition to learning about the technical aspects of cheesemaking, both you and your employees should understand the fundamentals of disease transmission and sanitation. According to the FDA, minimum training must include a written policy covering GMPs, personal hygiene, plant sanitation policies and procedures, food-safety and quality-control policies, and product-tampering awareness and consequences. Training must be presented in a language that can be understood by all employees. Training programs should be updated annually and records should be kept of training sessions. All new employees must be provided with initial training that covers the minimum requirements, and refresher courses should be provided quarterly. Operational deficiencies should result in additional training.

Food-Safety Strategy No. 2: Design Your Plant to Implement Good Manufacturing Practices (GMPs)

The construction and design of your facility should decrease the likelihood of food contamination. Proper design will avoid cross-over of flow paths of raw and finished products. It will also reduce the risk of food ingredients or finished food product contacting walls or floors and should prevent poorly drained floors. Plant design and construction should allow for proper equipment placement and adequate space around equipment, so that there are no barriers to

proper cleaning and sanitation of either food-processing equipment or facilities. Using footbaths, or better yet, chemical foamers, to control potential transfer of contaminants into the plant will be an important part of this plan. Remember that maintaining a clean footbath with the proper sanitizer at the correct dilution, as well as testing it to be sure it is effective, is the correct way to implement the use of a footbath.

Food-Safety Strategy No. 3: Follow Good Manufacturing Practices for Food Every Day

GMP food-safety strategy

From receiving of materials, process control, and finished product evaluation, all the way to product storage and shipping and employee training, consistent processes will mean a quality product. Good manufacturing practices aren't just a good idea—they are the law. These government regulations come from the Food, Drug and Cosmetic Act, Code of Federal Regulations: Title 21, Part 110 – Good Manufacturing Practices.

Title 21, Part 110 covers current good manufacturing practices in manufacturing, packing, or holding human food. These regulations cover the definitions of what is a good idea (should) and what is required (shall). Title 21, Part 110 broadly defines how personnel, buildings and facilities, sanitary operations, equipment, production and process controls, and defect action levels are to be managed. Some foods, even when produced under current good manufacturing practice, contain natural or unavoidable defects that at low levels are not hazardous to health. The Food and Drug Administration establishes maximum levels for these defects in foods produced under current good manufacturing practice. The definitions are broad enough to allow a baker and a cheesemaker to adapt the rules to fit their processes. The goal of all this is to insure that the cheese you make has been manufactured, prepared, packed, and held under sanitary conditions.

GMP highlights

Section 110.3. Definitions

Title 21, Part 110 contains 90 "Shalls" and 13 "Shoulds." This means the policy tells you a lot about what you must do and gives you some leeway as to what is a good idea, but it does not mandate exactly how to do it. This section also defines the following terms: *critical control point, food, food-contact surfaces, lot, microorganisms, pest, plant, quality-control operation, rework, safe moisture level, sanitize, and water activity.*

Section 110.5. Current good manufacturing practices for food: Criteria for determining adulteration

This section defines the criteria that determine whether a food is adulterated; has been manufactured under conditions that make it unfit to be classified as food; or has been prepared, packed, or held under unsanitary conditions so that it may have become contaminated with filth or injurious to a person's health.

Pest control outside the plant is important to prevent problems from getting inside. *Photo: Mike Poe.*

Part 110.10. Personnel

GMPs covering personnel discuss disease control, cleanliness, education and training, and supervision of personnel with regard to these requirements. This section includes information on hand washing, clothing, outer garments, hair, jewelry, and employee use of food, tobacco, or gum in the plant. It also covers employee illness.

Part 110.20. Plant and grounds

Grounds

A food manufacturer needs to ensure cleanliness and order both inside and outside the plant. This includes attending to drainage, waste disposal, pest harborage, and the minimizing of problems from adjacent property.

Plant design and construction

Addresses the placement of equipment and storage of materials to reduce potential contamination of food, food-contact surfaces, or food-packaging materials with microorganisms, chemicals, filth, or other extraneous material. This section also covers pest control, lighting, ventilation, and process flow.

Part 110.35. Sanitary operations

Discusses general maintenance issues in the plant, the range of substances that should be used in cleaning and sanitizing, the labeling and storage of toxic materials, pest control, the sanitation of food-contact surfaces, and the storage and handling of cleaned, portable equipment and utensils.

Part 110.37. Sanitary facilities and controls

Covers water supply and supply connections, plumbing, sewage disposal, toilet facilities, hand-washing facilities, and rubbish disposal. It also covers instructions to employees about sanitation and handling unprotected food.

Part 110.4. Equipment and utensils

Describes what materials are appropriate for food equipment and utensils, including mechanical equipment in the plant. It also covers conveyor systems, freezing and cold storage, instruments and controls used for measuring, regulating, or recording temperatures, pH, acidity, water activity, or other conditions that control or prevent the growth of undesirable microorganisms in food. This section also discusses the use of compressed air or other gases.

Part 110.8. Processes and controls

Addresses all operations in the receiving, inspecting, transporting, segregating, preparing, manufacturing, packaging, and storing of food. Raw materials and other ingredients have to be clean and suitable for processing into food, and they have to be stored under conditions that will protect against contamination and deterioration. Raw materials also need to be free of microbial or natural toxins that would cause them to be "adulterated."

Part 110.93. Warehousing and distribution

Requires that conditions for the storage and transportation of finished food shall protect food against physical, chemical, and microbial contamination and prevent deterioration of the food and the container.

Part 110.110. Natural or unavoidable defects in food for human use that present no health hazard

To read the complete text of Title 21, Part 110, you can go to the U.S Government Printing Office Web site (http://www.access.gpo.gov/) and download the law in PDF format.

Food-Safety Strategy No. 4: Develop and Follow a Master Sanitation Plan

Develop and follow a master sanitation plan that includes sanitation standard operating procedures (SSOPs). These are written procedures that are implemented in your processing plant. The goal of writing down and implementing the procedures is to prevent direct contamination or adulteration of your product. Even if you are operating your business on your own, a sanitation plan is a great reminder checklist, and the documentation could help you track down a potential problem.

Air blowers can greatly reduce the possibility of flying insects entering the plant. *Photo: Mike Poe.*

Pest traps are usually placed along room perimeters where rodents travel. *Photo: Mike Poe.*

How do SSOPs differ from GMPs? Good manufacturing practices (Safety Strategy No. 3) describe what you are required to do by law. SSOPs describe how procedures are done in your plant, and they are a prerequisite for a Hazard Analysis and Critical Control Point (HACCP) program. You develop your own SSOPs to fit your plant and manufacturing process. Unless you plan to work by yourself, SSOPs provide a set of instructions for your employees that tell them how every element of plant sanitation should be carried out.

HACCP programs involve a systematic, preventive approach to food safety in which the actions are taken at critical control points to prevent problems. We won't address HACCP programs here except to say that you must have good manufacturing practices in place and implement your SSOPs before you can consider trying to implement a HACCP program in your plant.

Early on, when you are planning your manufacturing process, you should write down the procedures you expect everyone to follow. By building SSOPs into the everyday practices of the plant, plant sanitation will be routine and you will create a "culture" of food safety.

Each SSOP procedure is broken down into the following parts:

- control measures

- monitoring procedures

- corrective actions

- recordkeeping

Your SSOPs cover the following elements of your processing:

- water quality

  » safety of the water that comes into contact with food or food-contact

surfaces or is used in the manufacture of ice

- food-contact surfaces

  » condition and cleanliness of food-contact surfaces, including utensils, gloves, and outer garments

- cross-contamination of unsanitary objects to the following:

  » food

  » food-packaging material

  » other food-contact surfaces

  » utensils

  » gloves

  » outer garments

  » raw milk to pasteurized product and finished product

- maintenance of toilets and hand-washing facilities

  » maintenance of facilities

  » hand washing

  » hand sanitizing

  » toilets

- product adulteration

  » protection of

    › food

    › food-packaging material

    › food-contact surfaces

  » from contamination by

    › chemical, physical, and biological contaminants

    › lubricants

    › fuel

    › pesticides

    › cleaning compounds

- › sanitizing agents
- › condensate
- toxic compounds
  - » labeling: clearly identified, in original container
  - » storage: away from food ingredients and food-processing areas
  - » use: avoid when there is a chance of cross-contamination
- employee health
  - » Prevent employees from contaminating food, clean equipment and utensils, and unwrapped single-service or single-use articles.
  - » Ask employees to report transmissible diseases.
  - » Restrict sick workers from normal work schedule in food-handling areas.
- pest exclusion
  - » ideal goal: pest-free conditions

You can see that the list above is almost exactly the same as the list of good manufacturing practices. They are meant to be parallel.

Below we have provided three examples for you: one list of questions that deals with water quality, one that covers cleaning and sanitation, and then a comprehensive example so you can get an idea of how to write an SSOP. Remember, in order to create any SSOP you would need to be able to answer all these questions, keep a written record of the questions and responses, and then follow your own recommendations all the time. Your SSOP must be designed around what you actually do in your manufacturing practices; you can't just use a template.

Remember, each SSOP covers the following:

- control measures
- monitoring procedures
- corrective actions
- recordkeeping

Sample water SSOP

Control measures

- What is the source of the water?
- Is it tested?
- Who performs maintenance and repair on the plumbing system?
- Does all work meet applicable codes?
- Are protective devices (anti-siphoning) installed as needed?
- Are floors sloped to facilitate drainage?

Monitoring

- Who does the water-quality monitoring?
- When is the monitoring done?
- What are the scheduled times?
- What special circumstances call for additional monitoring?
- What tests are carried out as part of the monitoring?
- Do the tests meet state and federal standards?
- At what locations within the plant is the testing done?
- If test results come back positive, is there any follow-up testing?

- What visual inspections are done on equipment or drains, and what is the frequency of inspection?

Corrective actions

- If the water system fails, what do you do?

- Under what conditions do you resume production?

- What do you do with product that was made during the interruption in normal service?

Recordkeeping

- Maintain reports for all testing, both for in-plant and water source.

- Record visual inspections.

- Document corrective actions.

- Maintain daily records sufficient to document the implementation and monitoring of the SSOPs and any corrective action taken.

- Specify the frequency with which each procedure in the SSOPs is to be conducted.

- Identify the employee(s) responsible for the implementation and maintenance of procedure(s).

Sample cleaning SSOP

Control measures

- Food-contact surfaces are to be adequately cleanable.

- Food-contact surfaces are to be cleaned and sanitized each day at the end of processing.

- Cleaning is to be done with the proper materials to clean and sanitize all food-preparation surfaces and all food-preparation areas.

- Workers are to wear clean gloves, waterproof aprons, and waterproof boots.

- Waterproof aprons are to be cleaned and sanitized at the end of the shift.

- Administrative personnel are to wear smocks and waterproof boots when in processing areas. Smocks are to be laundered in-house as needed.

Proper worker clothing and a clean, well-lit area for packaging your cheese helps ensure that a high-quality product will reach your customers. *Photo:* Steve Quirt.

- Maintenance workers are to wear uniforms and waterproof boots that distinguish them from food workers and their uniforms are to be laundered in-house as needed.

- The cleaning process for your cheese room and equipment is to be written down as a daily sanitation schedule, and it is part of a master sanitation schedule for all cleaning.

- The processing area is to be set up so that every surface can be cleaned—including under, behind, and above what you can easily see.

- Employees are to be trained in how to do the cleaning.

- The chemicals used are to be labeled for their use in cleaning, and label directions are to be followed.

- There is to be a designated place for chemicals to be stored and mixed.

  Be sure employees know what equipment is to be cleaned in place (CIP) and what equipment is to be disassembled and cleaned in a sink (cleaning out of place or COP), as well as what environmental cleaning is to be done. They should know the specific amount of time required for cleaning as well as the water temperature and the correct cleaning agents to be used.

Monitoring

A specific employee or manager is designated to

- inspect food-contact surfaces to determine if they are adequately cleanable and are clean and sanitized before processing begins and after each cleanup period

- monitor the use of gloves and the cleanliness of workers' outer garments

- record that cleaning has been completed

- inspect the cleaning to be sure it is done properly

- test chemicals when mixed to be sure they are at the proper concentrations

- conduct confirmatory testing (environmental swabs) to be sure the cleaning process is effective

- ensure that all employees are trained to change aprons, coats, and gloves as they become soiled or wet

Corrective actions

- Are food-contact surfaces that are not adequately cleanable repaired or replaced?

- Are food-contact surfaces that are not clean recleaned?

- Are gloves that become a potential source of contamination cleaned and sanitized or replaced? Are outer garments that become a potential source of contamination cleaned and sanitized or replaced?

- If a cleaning procedure is not complete or is not followed, what are the corrective actions that are taken?

- Are employees retrained as needed?

Recordkeeping

- Is the condition of food-contact surfaces, sanitation inspections, use and cleanliness of gloves, cleanliness of worker outer garments, and corrective actions noted on the daily sanitation report?

- Do you keep a daily sanitation report and a master sanitation schedule?

- Where do you record visual inspections?

- Where do you record corrective actions?

Model sanitation standard operating procedures

The following model SSOP addresses the sanitation concerns for a fictional cheese company. SSOPs will vary from facility to facility because each facility and process is designed differently. This SSOP is for illustrative purposes. The use of trade names does not constitute endorsement for any specific product.

> ABC Cheese Company
> 1 Whey Lane
> Cheeseville, Any State

1. Safety of the Water That Comes into Contact with Food or Food-Contact Surfaces or Is Used in the Manufacture of Ice

Control measures
All water used in the plant is from a reliable municipal water system. The water system in the plant was designed and installed by a licensed plumbing contractor, and it meets current community-building and food-manufacturing codes. All modifications to the plumbing system will be completed by a licensed plumbing contractor and will be inspected to ensure conformance with codes. All hoses inside and outside the plant have anti-siphoning devices installed. Floors are sloped to facilitate drainage.

Monitoring procedures
The municipal water district routinely monitors the water to ensure that it meets state and federal water-quality standards. The quality-assurance supervisor receives and reviews annual reports of municipal water quality.

Twice a year, and when modifications are made to the plumbing system, water samples from at least four locations in the plant are sent to a private testing laboratory and examined for the presence of coliforms. Cultures testing positive for coliforms are examined for the presence of fecal coliforms. The quality-assurance supervisor receives and reviews the laboratory reports.

Hoses are inspected daily during production to ensure the presence of functional anti-siphoning devices. Floors in the processing area are inspected daily during production for adequate drainage.

Corrective actions
In the event of municipal water supply failure, the plant will stop production, determine when the failure occurred, and embargo all products produced during the failure until product safety can be assured. Production will resume only when water meets state and federal water-quality standards.

If in-plant sampling indicates the presence of coliform bacteria in more than 5 percent of plant water samples, the plant will contact the municipal water system and inspect the plumbing system to determine the source of the coliform bacteria. Corrections will be made to the plumbing system or well, if necessary, to correct problems.

If in-plant sampling indicates the presence of fecal coliforms in any plant water sample, the plant will stop production and embargo all products until product safety can be assured. The plant will contact the municipal water system and inspect the plumbing system to determine the source of the fecal coliforms. Corrections will be made to the plumbing system, if necessary, to correct problems. Production will resume only when water meets state and federal water-quality standards.

Hoses without anti-siphoning devices will be red-tagged and will not be used until anti-siphoning devices have been installed. Floors with standing water will have the drains unplugged, or, if necessary, consultations will be held with plumbing or general contractors and modifications will be made to correct floor drainage problems.

Recordkeeping

Reports are kept for municipal water quality, in-plant water-quality testing, and corrective actions. Hose inspections, floor drainage inspections, and corrective actions are recorded on the daily sanitation report.

2. Condition and Cleanliness of Food-Contact Surfaces, Including Utensils, Gloves, and Outer Garments

Control measures

- Food-contact surfaces are adequately cleanable.

- Food-contact surfaces are cleaned and sanitized each day at the end of processing.

- In the morning before processing: food-contact surfaces are rinsed with cold water and sanitized with a 100-ppm sodium hypochlorite sanitizer.

- At the end of manufacturing: major solids are physically removed from floors, equipment, and food-contact surfaces. Equipment is disassembled as required for adequate cleaning. All surfaces are rinsed with cold water. Equipment and food-contact surfaces are scrubbed using brushes with a chlorinated alkaline cleaner in warm (120°F) water. All surfaces and floors are rinsed with cold water. Floors and walls are sprayed with a 100-ppm sodium hypochlorite sanitizer

solution. Utensils are cleaned in a deep sink with a chlorinated alkaline cleaner, rinsed in hot water (190°F), soaked in a 100-ppm sodium hypochlorite sanitizer for at least 10 minutes, and air-dried.

- Clothing and protective wear: employees working with milk and cheese products wear clean gloves, waterproof aprons, and waterproof boots. Waterproof aprons are cleaned and sanitized at the end of the shift. Administrative personnel wear smocks and waterproof boots when in processing areas. Smocks are laundered in-house as needed. Maintenance workers wear gray uniforms and waterproof boots. Uniforms are laundered in-house as needed.

For information about organic standards and a list of allowed substances for cleaning, refer to the USDA's Web site on the National Organic Program at www.ams.usda.gov/nop/.

Monitoring procedures

The sanitation supervisor inspects food-contact surfaces to determine if they are adequately cleanable and are clean and sanitized before processing begins and after each cleanup period. Production supervisors monitor the use of gloves and the cleanliness of workers' outer garments.

Corrective actions

Food-contact surfaces that are not adequately cleanable are repaired or replaced. Food-contact surfaces that are not clean are recleaned. Gloves that become a potential source of contamination are cleaned and sanitized or replaced. Outer garments that become a potential source of contamination are cleaned and sanitized or replaced.

Recordkeeping

Condition of food-contact surfaces,

sanitation inspections, use and cleanliness of gloves, cleanliness of workers' outer garments, and corrective actions are noted on the daily sanitation report.

3. Prevention of Cross-Contamination from Insanitary Objects to Food, Food-Packaging Material, and Other Food-Contact Surfaces, Including Utensils, Gloves, and Outer Garments, and from Raw Product to Cooked Product

Control measures

Production supervisors have received basic food-sanitation training. Workers wear hairnets, headbands, caps, beard covers, or other effective hair restraints and do not wear jewelry or other objects that might fall into the product, equipment, or containers. Workers wear disposable gloves and replace them as needed. Workers wash their hands and gloves thoroughly and sanitize them before starting work, after each absence from their workstation, and any time they have become soiled or contaminated.

Clothing and personal belongings are not stored in production areas. Workers do not eat food, chew gum, drink beverages, or use tobacco in production areas. Except for cheesemakers, no one is allowed to enter or pass through other processing areas during processing.

Workers sanitize their boots in boot baths containing 800-ppm hypochlorite sanitizer solution before entering processing areas. Boot sanitizing solutions are tested daily and changed as needed. (Although quaternary compounds are commonly used for footbaths, they are not recommended for sanitation in cheese plants because they can interfere with cheese cultures.)

Cleaning and sanitizing equipment is color-coded for specific plant areas: blue for raw-product processing areas, white for cooked-product processing areas, and yellow for toilet facilities and general plant cleaning.

Plant grounds are in a condition that protects against contamination of food. Waste is removed from processing areas every 4 hours during production.

Plant buildings are maintained in good repair. Milk-holding and cheese-processing areas are separated. Drip or condensate does not contaminate food or packaging materials. Safety-type light fixtures are used in processing and packaging areas. Coolers, including the evaporators, are cleaned annually or more often if needed. Nonfood-contact surfaces in processing and packaging areas are cleaned daily at the end of the shift.

Milk supplies for cheesemaking and cheese in the process of being made is physically separated from finished product. Packaging materials are protected from contamination during storage.

Monitoring procedures

Plant manager schedules basic food-sanitation courses for new production supervisors.

Production supervisors monitor hair restraint use, glove use, hand washing, personal belonging storage, eating and drinking in processing areas, and boot sanitizing.

Sanitation supervisor monitors use of proper sanitation equipment and removal of waste from processing areas.

Sanitation supervisor inspects the plant and coolers before processing begins. Sanitation supervisor inspects packaging material storage area and plant grounds daily.

Corrective actions

- Production supervisors receive basic sanitation instruction.

- Workers correct deficiencies in hair restraint use, glove use, hand washing, personal belonging storage, eating and drinking in processing areas, and boot sanitizing before working in milk storage or cheese-processing areas.

- Sanitation equipment that is being used in the wrong plant area is cleaned and sanitized and exchanged for correct equipment. Sanitation supervisor initiates correction of any potentially contaminating condition.

- Milk and unfinished cheeses are physically separated from finished products, and potentially contaminated finished products are destroyed.

Recordkeeping

- Training records indicate that production supervisors have received basic food-sanitation training.

- Hair restraint use, glove use, hand washing, personal belonging storage, eating and drinking in processing areas, boot sanitizing, use of proper sanitation equipment, plant grounds and waste inspections, plant and cooler inspections, packaging material storage inspections, and corrective actions are noted on the daily sanitation report.

4. Maintenance of Hand-Washing, Hand-Sanitizing, and Toilet Facilities

Control measures

Toilet facilities are provided adjacent to but physically separated from processing areas. Toilet facilities have self-closing doors, are maintained in good repair, and are cleaned and sanitized daily at the end of the shift.

Hand-washing facilities are provided in milk-holding and cheese-processing areas and in the toilet facility. Toilet facilities and hand-washing stations have the following:

- hot and cold running water with foot-activated valves

- liquid sanitizing hand soap

- hand sanitizer solutions that are tested every 4 hours during production and changed as needed

- sanitary towel service

- signs directing workers to wash their hands and gloves thoroughly and sanitize them before starting work, after each absence from their workstation, and any time they have become soiled or contaminated or have come in contact with refuse receptacles

Monitoring procedures

- Sanitation supervisor inspects the toilet facilities and hand-washing facilities daily.

Corrective actions

- Sanitation supervisor initiates cleaning of dirty toilet facilities and correction of any potentially contaminating condition. Repairs are made as needed.

Recordkeeping

- Inspections of toilet and hand-washing facilities, as well as corrective actions, are noted on the daily sanitation report.

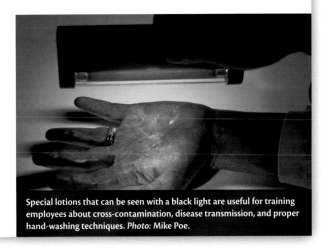

Special lotions that can be seen with a black light are useful for training employees about cross-contamination, disease transmission, and proper hand-washing techniques. *Photo: Mike Poe.*

5. Protection of Food, Food-Packaging Material, and Food-Contact Surfaces from Adulteration with Lubricants, Fuel, Pesticides, Cleaning Compounds, Sanitizing Agents, Condensate, and Other Chemical, Physical, and Biological Contaminants

Control measures

Cleaning compounds, sanitizers, and lubricants used in processing and packaging areas are listed in the USDA's Food Safety and Inspection Service's "List of Proprietary Substances and Nonfood Compounds Authorized for Use under USDA Inspection and Grading Programs" (FSIS Miscellaneous Publication No. 1419). Food-grade and nonfood-grade chemicals and lubricants are stored separately outside processing and packaging areas. Food, food-packaging materials, and food-contact surfaces are protected from adulteration from biological, chemical, and physical contaminants.

Monitoring procedures

Invoices are checked at receiving before chemicals are stored in the food-grade chemical storage area. Sanitation supervisor inspects chemical storage areas daily and inspects processing and packaging areas daily before production begins.

Corrective actions

Unapproved chemicals are returned or used in nonprocessing areas. Improperly stored chemicals are moved to the correct storage area. Sanitation supervisor initiates correction of any potentially contaminating condition. Repairs are made as needed.

Recordkeeping

Invoices are kept for food-grade chemicals and lubricants. Chemical storage area, processing and packaging area inspections, and corrective actions are noted on the daily sanitation report.

6. Labeling, Storage, and Use of Toxic Compounds

Control measures

Cleaning compounds, sanitizing agents, lubricants, and pesticide chemicals are properly labeled. They are stored outside processing and packaging areas and separately from packaging materials. Food-grade chemicals and lubricants are stored separately from nonfood-grade chemicals and lubricants.

Monitoring procedures

- Sanitation supervisor inspects chemical storage areas daily.

Corrective actions

- Unlabeled chemicals are removed from storage areas and properly disposed of. Improperly stored chemicals are moved to correct storage areas.

Recordkeeping

- Chemical storage area inspections and corrective actions are noted on the daily sanitation report. Material safety data sheets (MSDS) are kept for all nonfood-grade chemicals.

7. Control of Employee Health Conditions That Could Result in the Microbiological Contamination of Food, Food-Packaging Materials, and Food-Contact Surfaces

Control measures

- Workers are instructed to report to their immediate supervisor any health condition that might result in food contamination.

Monitoring procedures

- Supervisors report suspected health

problems to the plant manager. The plant manager decides if a potential food contamination situation exists.

Corrective actions

- Workers who represent a potential risk to food safety are sent home or reassigned to nonfood-contact jobs.

Recordkeeping

- Worker health and corrective actions are noted on the daily sanitation report.

8. Exclusion of Pests from the Food Plant

Control measures

A pest management firm treats the outside of the building every other month. They also inspect the interior of the building and implement appropriate pest control. Plant grounds and interior areas are kept free of litter, waste, and other conditions that might attract pests. Outer doors to the building are kept closed, processing areas are screened with plastic curtains, and electric bug-killing devices are located outside entrances to processing areas. No pets are allowed in the plant. Supervisors report any pest problems to the plant manager.

Monitoring procedures

- The plant manager reviews reports of pest treatment. The sanitation supervisor inspects the plant's exterior and interior daily.

Corrective actions

- The pest management firm is notified of any pest problem and treats the problem. Pest treatments are more frequent if problems are identified.

Recordkeeping

- Records of pest treatment: plant inspections and corrective actions are noted on the daily sanitation report.

- Reasons for creating a daily sanitation report: you can be sure that things are working smoothly at your facility and that you are satisfied with conditions for food processing. Any problems can be noted and remedied in a short time frame. With proper observation and actions, things will stay running smoothly. Figure 6.1 is a sample daily sanitation report that can be adapted to your cheesemaking process. Note that it is used before, during, and after food processing.

- The master sanitation schedule, shown in figure 6.2, establishes a routine for cleaning. Remember that small, routine steps can prevent big problems from going unnoticed.

Food-Safety Strategy No. 5: Conduct Regular Environmental Swabbing

Conduct regular environmental swabbing in your plant to verify the effectiveness of your sanitation practices. This procedure goes hand in hand with good manufacturing practices and sanitation standard operating procedures.

Environmental swabbing can be done in two ways:

Nonspecific swabbing

Nonspecific swabs can determine if any biological material remains on a surface that should be sanitary. The swabs are designed to produce a fluorescent color that can be read in a small monitor. The color is correlated to the amount of adenosine triphosphate (ATP) detected on the swabbed

Figure 6.1. Example of daily sanitation report.

Daily Sanitation Report — ABC Cheese, Inc.	Date:		
Condition	**Initials (passes inspection)**		
	Before processing	**Midday cleanup**	**End-of-shift cleanup**
Plant grounds do not cause food contamination.			
Waste is properly stored.			
Equipment and utensils are adequately cleanable.			
Food-contact surfaces and utensils are clean and sanitized.			
Food, food-contact surfaces, and packaging materials are protected from adulteration/contaminants.			
Nonfood-contact surfaces are clean.			
Hoses have anti-siphoning devices. Floors have adequate drainage.			
Coolers and evaporators are clean.			
Finished and unfinished products are physically separated in coolers.			
Toilet facilities are clean, sanitary, and in good repair.			
Toxic compounds are identified and stored properly.			
Employee health conditions are acceptable.			
Gloves and garments contacting food are clean and sanitary.			
Employee practices do not result in food contamination (hair restraints, glove use, hand washing, personal belonging storage, eating and drinking, boot sanitizing).			
Proper color-coded sanitation equipment is used.			
Hand and boot sanitizer strength is adequate.			
No pests are in the plant.			
Deviations from SSOP and corrective actions:			
Reviewed by (plant manager):		**Date:**	

Note: Shaded areas of the checklist need no action.
Source: Adapted from the Seafood Network Information Center (SeafoodNIC), a part of the University of California Sea Grant Extension Program.

Figure 6.2. Master sanitation schedule.

Master Sanitation Schedule

Big Cheese Company
Small Cooler

Month of: _____

Cleaning to be completed	DATE - WEEK ONE								
	week one	week one	week one	week one	week one				
Clean ceilings									
Sweep perimeters					initial	initial			
Clean cheese racks		initial	initial						

Cleaning to be completed	DATE - WEEK TWO								
	week two	week two	week two	week two	week two				
Clean ceilings									
Sweep perimeters					initial	initial			
Clean cheese racks		initial	initial						

Cleaning to be completed	DATE - WEEK THREE								
	week three	week three	week three	week three	week three				
Clean ceilings									
Sweep perimeters					initial	initial			
Clean cheese racks		initial	initial						

Cleaning to be completed	DATE - WEEK FOUR								
	week four	week four	week four	week four	week four				
Clean ceilings									
Sweep perimeters					initial	initial			
Clean cheese racks		initial	initial						

Reviewed by _____

Note: Shaded areas of the checklist need no action.

surface. ATP is produced by almost all living things and, if found, indicates that the surface is not free of biological material. This could be food residue from incomplete cleaning or microbial activity from incomplete sanitation. These tests take only a minute and give immediate feedback to monitor compliance with your sanitation and cleaning programs.

Microbe-specific swabbing

In cheese plants that have not been cleaned properly or that have problems with the introduction of microbes from outside the plant, *Listeria* spp. can become a serious problem. Even if *Listeria* spp. is not on a food-production surface, the fact that it is in the plant increases the risk of product contamination. You need to focus on *Listeria* for environmental swabbing because it is an organism that is widespread in the environment and it grows under refrigerated conditions. It is killed by heat, and sanitation is very important.

When swabbing for *Listeria*, work with a recognized analytical laboratory to develop a swabbing protocol and obtain swabs. The lab should provide you with instructions on the use and return of the swabs so they can complete the swab cultures back at their laboratory.

You may ask yourself: Why should I do all this testing? Why not just take samples of product to prove there isn't a problem? The statistical reality is that you cannot take a large enough sample of your product to "prove" that it is free of bacterial contamination. In order to get a large enough sample, you would have nothing left to sell.

An example of the challenges of product sampling is given in figure 6.3. By focusing your efforts on the potential effectiveness of your cleaning and sanitation programs, you can decrease the risk of product contamination.

Food-Safety Strategy No. 6: Develop and Follow a Pest Management Plan

If necessary, hire an outside firm to implement a pest management plan for you. One of the most challenging issues for new business owners is keeping up with all the tasks they have. In most cases, something is left undone, even if priorities have been set. There just aren't enough hours in the day.

Pest management is one area that takes routine time in the schedule, and it may also require special licensing and continuing education in order to be legal. This is because pest control (especially outside the plant) may require the application of chemicals that require a permit.

The best defense against pests is to be sure they don't have access into your premises. The building should be "pest proof," and you should have exclusion strategies to prevent them from entering the building. This includes self-closing doors, screens, a well-constructed building in which all openings for ventilation, cables, and pipes are screened or caulked or otherwise sealed to prevent pest entry.

You can contract with a pest management firm to spray for insects outside of the building and inspect the interior of the building, controlling pests in compliance with GMPs. Please note that pesticides cannot be used in food-production areas. The company can set traps for rodents and other small vertebrate pests both inside and outside the building, check them routinely, and replace them as needed. The company can also install insect traps and monitor and empty them as needed.

Plant construction and maintenance, along with routine sanitation and cleaning,

Figure 6.3. Problems with end product testing.

Why product sampling is not a good way to "prove" that your cheese is not contaminated with a pathogen

Testing for pathogens in end products has a limited place in a comprehensive food-safety program. Microbial testing is a destructive process; samples that are analyzed are destroyed in the process. In addition, it is still a lengthy process. Even so-called rapid tests take 1 to 2 days, and often these results are only "presumptive" positive. Confirmation can take as much as a week. Pathogen testing is also best applied to products that are of uniform microbial composition, that can be easily and well mixed prior to sampling, and that have reasonably high levels of pathogens, should they be present. In products that are solid and that have undergone treatments designed to reduce or eliminate pathogens (such as cheese), these conditions do not apply.

The following example illustrates this point. Let's assume that ABC Cheese Company has a pathogen problem that results in 1 percent contamination of wheels of cheese with either *Listeria monocytogenes* or *Salmonella typhimurium* (an unusually high contamination level for a well-run cheese plant). So, in the aging room of 300 2.5-kilogram wheels, there are three wheels that are contaminated with *Listeria* (red) or *Salmonella* (black). The ABC Cheese Company only tests for *Listeria* and not *Salmonella*.

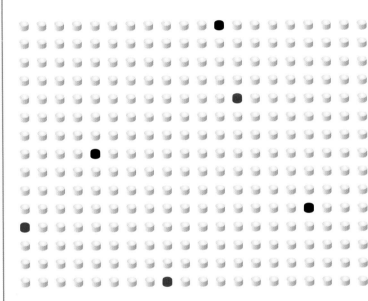

At left is the ABC cheese aging room with the 300 2.5-kilogram wheels representing one production lot. The ABC quality-assurance manager has a 1 in 100 chance of picking a wheel that contains the organism she is looking for (*Listeria*).

Let's assume that one of the positive wheels of cheese is selected. Below is a diagram which represents two contaminated wheels of cheese from the ABC Cheese Company. One of the 2.5-kilogram wheels is contaminated with *Listeria monocytogenes* (red) and the other with *Salmonella typhimurium* (black), but only in one 25-gram area. The bacteria are not evenly distributed throughout the wheel, which would be normal for this type of sporadic contamination. The diagram below shows what the wheel looks like from the perspective of the 25-gram samples. A 25-gram sample is a standard size for pathogen testing.

IF the ABC quality-assurance manager takes a single 25-gram sample out of the wheel that does have the organism she is looking for, she has a 1 in 100 chance of sampling any given wheel correctly. Even if she takes a 100-gram sample to mix before removing the 25-gram sample, she still only has a 1 in 25 probability of sampling the contaminated portion.

So, her overall odds of actually finding the contaminant she is concerned about without the 100-gram sample mixing are 1/100 X 1/100 = 0.0001, or 1 in 10,000. Or, with the 100-gram sample mixing, the odds are 1/100 X 1/25 = 1 in 2,500. Is this a level of confidence you can live with when it comes to assuring your customers of a safe and wholesome product? Even if the cheese wheel and subsample containing *Salmonella* were selected, it would not be detected using the methods for *Listeria*.

What if the cheese was contaminated with *E. coli* O157:H7 or any of the other foodborne pathogens? Even if the correct sample is selected, you can only find the organism you are testing for.

Let's also consider the reverse situation. If a positive is indeed found with the lot of cheese illustrated above, what does it tell us about contamination of the entire lot? Can subsequent retesting of the product assure that wheels of cheese are free of the pathogen?

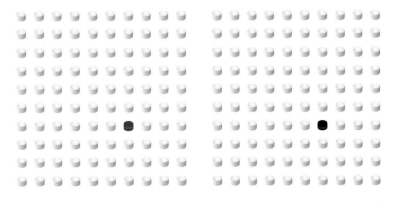

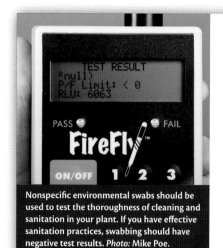

Nonspecific environmental swabs should be used to test the thoroughness of cleaning and sanitation in your plant. If you have effective sanitation practices, swabbing should have negative test results. *Photo: Mike Poe.*

Swabs are designed to detect adenosine triphosphate (ATP) from living cells found in organic matter. *Photo: Mike Poe.*

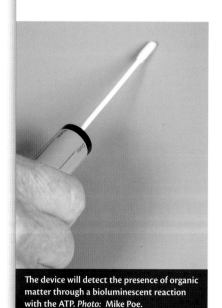

The device will detect the presence of organic matter through a bioluminescent reaction with the ATP. *Photo: Mike Poe.*

are important preventative measures for keeping pests out of the plant. This is an important job. You and your staff may have time to keep the grounds and interior areas free of litter, waste, and other conditions that might attract pests. You can also make sure that outside plant doors are kept closed and that processing areas are screened with plastic curtains.

Pets should never be allowed in the plant! Everyone involved in food processing should avoid contact with pets if they roam free outside the plant. Employees should also avoid contact with farm animals during their shift. This is especially important in a farmstead setting where people can track in contaminants from pets or other animal areas on the farm.

Food-Safety Strategy No. 7: Develop and Follow a Chemical Control Plan

An important element of food safety is preventing contamination of food through the appropriate use, storage, and labeling of all chemicals used in the processing facility. This includes laboratory supplies, food-grade materials like starter cultures and rennet, as well as toxic chemicals such as soap, acid, and sanitizers. In addition, proper cleaning and sanitation can only be accomplished when chemicals have been diluted to the proper concentration, and when they have not lost their activity due to contact with organic matter.

When using chemicals for cleaning in a food-processing plant, it is essential that you mix them with potable water and that the solution is freshly mixed at the time of use. It is also essential that the solution is never reused and that the concentration is accurate.

If the concentration is too low, the product may not clean or sanitize properly—and if the concentration is too high, you will be violating the regulated use of the product.

Why are there different chemical concentrations? Chemicals have been designed to work under different conditions, including

- on heated surfaces

- on cold surfaces

- in clean-in-place systems (CIP)

- in clean-out-of-place tanks (COP)

Each chemical you purchase should come with a test kit.

Each test kit should tell you

- what chemical you are checking

- how to titrate each chemical

- how many parts per million are in each drop

Rules for working with chemicals include the following:

- When refilling or carrying concentrated chemicals, you must have a closed container.

- Wear your protective equipment, including face shield or glasses, gloves, and an apron.

- Never mix different chemicals.

- Label all chemical containers properly.

- Be sure you have a cap for the container.

- Be sure you are putting the same chemicals in the container each time.

- If there is a leak in the chemical container, replace the container.

What should you do if you have an accident? All the chemical products you use will come with a material safety data sheet (MSDS). According to OSHA, it is your responsibility to train all your personnel on what to do in the event of a chemical spill or other accident with chemicals, and to have the MSDS available.

Figure 6.4 highlights the difference between cleaning, sanitizing, and sterilizing surfaces. These terms are often used incorrectly when cheesemakers don't understand the difference between them. Table 6.2 shows a sample record of cleaning and lubricating products as well as testing and dilution rates.

Food-Safety Strategy No. 8: Establish an Allergens Protocol

Establish a protocol for dealing with ingredients considered to be allergens. Your protocol will include your product-labeling practices. The eight common allergens are milk, eggs, peanuts, tree nuts (such as almonds, cashews, and walnuts), fish (such as bass, cod, and flounder), shellfish (such as crab, lobster, and shrimp), soybeans, and wheat.

Ingredient master list
Develop a master list of all ingredients in your facility, including spices, flavorings, additives, cultures, and minerals (like salt), and specify those that are allergens or that contain allergens. Your list should state which finished products contain allergenic ingredients.

Suppliers and raw materials
Ask your suppliers if they use ingredients that are allergenic. Require them to have

Figure 6.4. Definition of key cleaning terminology.

Increasing Degree of Microbial Kill

Food-processing industry	Hospital	Medical industry
sanitizer/sanitize	**disinfectant/disinfect**	**sterilant/sterilize**
Reduces microbial contamination to a safe level. Two types: 1) no-rinse food-contact surface sanitizer 2) nonfood-contact surface sanitizer	Kills 100% of vegetative cells. May not kill bacteria spores.	Completely destroys all forms of life. Includes bacterial spores as well as bacteria, viruses, and fungi. Heat and some chemicals effective under specific conditions.

Table 6.2. Sample of cleaning products, with their strengths and dilution rates to be followed by employees

Chemical (names are fictitious)	Strength	Dilution
Chlorinated alkaline cleaner **Brand:** Super Kleen **Usage:** equipment, food-contact surfaces, utensils, toilet facilities		1 ounce of concentrate to 3 gallons of water
Liquid sanitizing hand soap **Brand:** Eco Friend M3 **Usage:** hand-washing facilities		undiluted
Sodium hypochlorite sanitizer **Brand:** Eco Friend Sani-Special **Usage:** Food-contact surfaces, utensils	100 ppm	1 ounce of concentrate to 6.5 gallons of water
Sodium hypochlorite sanitizer **Brand:** Eco Friend Kill Bac **Usage:** Floors **Usage:** Boot-sanitizing baths	400 ppm 800 ppm	1 ounce of concentrate to 2 gallons of water 1 ounce of concentrate to 1 gallon of water
Iodine sanitizer **Brand:** Super Kleen **Usage:** Hand-sanitizing solutions	25 ppm	1 ounce of concentrate to 13 gallons of water
Lubricants **Brand:** Greasy Industries Special Super Grease (H-1) **Usage:** food-processing equipment **Brand:** Greasy Industries White Grease Slick Special **Usage:** nonfood-processing areas	not applicable	full strength

a documented allergen control program in place, and to provide you a letter with each shipment guaranteeing that purchased ingredients are free of undeclared allergens.

Receiving and storage

Allergens should be shipped in clearly marked, sealed containers and be kept physically separated from nonallergens. All shipments should be inspected for damaged containers and spillage.

Ideally, allergenic ingredients should be clearly marked as allergens and isolated from allergen-free products in storage. If space is limited, keep 4 feet between allergens and allergen-free products. Allergenic ingredients should also be stored below nonallergen products.

Production and scheduling

Dedicate processing equipment, personnel, and production lines to allergenic products to prevent cross contact. If this is not an option in your plant, manage allergens with production scheduling.

- Schedule long runs of products containing allergens to minimize changeover.

- Segregate production so that allergen-containing product is produced on separate days of the week.

- Run allergen-free products before allergen-containing products.

- Clean the manufacturing area immediately following the production of allergenic products to help reduce the risk of allergenic residue being transferred to new products.

- Test equipment-contact surfaces to detect the presence of residual allergens.

Labeling and packaging

Make sure correct packaging materials and labeling are used. Be sure that labels for allergen-containing products are not mixed with allergen-free food labels. Check label accuracy against the product's declared ingredients.

The FDA has published labeling compliance guidelines for manufacturers. Use plain language on all labels. For example: "This product contains milk." Product changes should be immediately reflected on labels.

Cleaning and sanitation

To comply with GMPs, be sure that your sanitation schedules and SSOPs are followed and documented. To help fulfill the GMPs, crews should disassemble equipment as necessary and focus on hard-to-clean areas to be sure cleaning is complete.

Verification tools

Work with independent third-parties to help you successfully implement your program and verify its effectiveness through auditing, testing, and employee training.

Auditing

It is important to audit your suppliers on a regular basis to assess the effectiveness of their allergen control program. In addition, you should audit your facilities to be sure employees are in compliance with your internal allergen program.

Testing

Testing results provide benchmarks for monitoring the effectiveness of cleaning and sanitizing procedures, which helps you prevent cross-contamination. Analytical laboratories can help you develop a compre-hensive testing program to pinpoint problem areas so you can institute corrective actions. Commercial test kits using ELISA (enzyme-linked immunosorbent assay) technology are also readily available to manufacturers.

Employee training

Even traces of allergens can induce mild to severe allergic attacks in susceptible people. Communicate this to all your employees in routine training. Employees must understand the eight major allergens, financial ramifications of recalls, potential areas of cross contact in the plant environment, and the importance of accurately declaring ingredients on product labels.

Food-Safety Strategy No. 9: Establish a Customer and Consumer Complaints Protocol

If there are any problems with your product, you want to know about it as soon as possible. Figure 6.5 is a sample consumer complaint form.

As any business operator knows, some consumer complaints are frivolous or even false. However, any complaints linking your products to product defects or illness must be taken seriously.

Regulations may require you to include calorie and nutritional information on the cheese packaging. This will depend in part on the number of units you sell. *Photo: Steve Quirt.*

Figure 6.5. Sample customer complaint form.

ABC Cheese Company Customer Complaint Form

Person taking complaint: _____ Date of complaint: _____

Customer name: _____

Street address: _____

City, state zip:_____ Home telephone: _____

Specifics of complaint:

Product name: _____

Lot code: _____ Sell-by date: _____

Date purchased: _____ Location of purchase:_____

Reported problem with product: _____

_____

_____

_____

If the caller reports illness that he or she believes to be related to the product:

Symptoms: _____

_____

_____

Has the caller received medical attention or diagnosis? _____

If so, name and address of medical professional: _____

_____

_____

Has the caller contacted the local health department? _____

 If so, date and official contacted: _____

Resolution of this complaint:

What steps were taken to resolve the complaint? _____

_____

_____

_____

Person resolving complaint: _____

Today's date:_____

As the business owner, you may ask employees to notify you immediately of any complaint. At the very least, keep a written log of all complaints and make it a part of plant operations to review it on a regular basis. This will allow you to track problems. Here are some tips on handling customer complaints:

- Obtain the caller's name, address, and phone number.

- Write down the specifics of the complaint. Include details about problems with the product. If the person says he or she is sick, also document the symptoms experienced, when they occurred, what product the caller thinks caused the illness, where and when that product was purchased, and when the product was eaten.

- Ask the caller how the product was handled between purchase and consumption.

- Ask if the caller has received medical care or diagnosis, and whether he or she called the local health department.

- If you get calls from two or more unrelated customers who are ill and complaining about a particular product, call your local or state health department.

Food-Safety Strategy No. 10: Develop a Food Recall Plan

Develop a food recall plan and practice going through a mock recall to be sure the plan works the way you expect it to. What is a recall? A product recall is a firm's removal or correction of a marketed product.

The purpose of a product recall is to

- promote customer and consumer safety

- locate the product in question and remove it from the marketplace at minimum cost and inconvenience to the customer

- provide accurate information to customers when appropriate

- comply with applicable laws and government regulations

A company may find out about a problem in any of the following ways:

- Internal quality assurance testing indicates that there may be a potential problem with a particular product or batch (e.g., microbiological results are outside acceptable limits).

- Customer or wholesaler gives feedback or complaints (e.g., a phone call or letter informing sales department of a potential problem).

- Supplier of a raw material used by the company in making its products informs management that there is a problem with an ingredient.

- Government sends notification through entities such as the California Department of Food and Agriculture or the Centers for Disease Control in the event of a widespread foodborne disease outbreak traced back to a specific product. The 2011 Food Safety Modernization Act will request more details and deadlines for a HACCP and food recall plan as it goes into effect over the next several years.

What is a "normal" sequence of events in a recall procedure?

1. Receive complaint or QA/QC finds problem.

2. Assess public health significance.

3. Formally notify regulatory agency.

4. Determine classification and depth of problem.

5. Monitor the recall.

6. Terminate the recall.

Who says I need to do a recall? A recall can be voluntary or several agencies may be involved, including the state Department of Agriculture, the Food and Drug Administration, the Food Safety Inspection Service, and the U.S. Department of Agriculture.

Recall classifications

- **Class I recall:** This is a health-hazard situation where there is a reasonable probability that the use of the product will cause serious, adverse health consequences or death. Priority must be given to the immediate and complete removal of all suspected product from the marketplace.

- **Class II recall:** This is a health-hazard situation where there is a reasonable probability of adverse health consequences from the use of the product but it is not life threatening.

- **Class III recall:** This is a situation where the use of the product will not cause adverse health consequences.

  - » **Market withdrawal:** Voluntary removal or correction of a distributed product that involves a minor infraction that would not warrant legal action and is not a health hazard.

  - » **Stock recovery:** Removal or correction of product that has not been marketed or that has not left the direct control of the firm. These products have been mislabeled, poorly packaged, damaged in shipment, or are not up to the standards expected by the customer.

Depth of recalls

- **Consumer level:** This includes household consumers as well as all other levels of distribution.

- **Retail level:** This includes all retail sales of the recalled product.

- **User level:** This includes hotels, restaurants, and other foodservice institutional consignees.

- **Wholesale level:** This is the distribution level between the manufacturer and the retailer. This level may not be encountered in every recall situation. Table 6.3 contains product recall examples for cheeses.

In order to conduct a recall effectively, you need to have a plan in place before you ever have a problem, and you need to test your plan to see how effectively your plan works. Below is a sample recall plan that you can adapt to your own needs.

Table 6.3. Recall examples

Product	Problem	Distribution	Status
fresh Chèvre	High total viable count (TVC). Microbiological results indicate the presence of pathogens.	Product has been distributed at retail level.	Consumer level: recall product. Food poses a safety risk. Results indicate pasteurization was incomplete or post-production contamination occurred.
Quark	Wrongly labeled "best before" date. ("Best before" date was accidentally set too far ahead and not to product specification).	Product has been distributed at retail level.	Withdraw product. The "best before" date of the product is inaccurate due to mislabeling.
vacuum pack, random-weight Gouda	Defective or blown packaging (high TVC) or faulty vacuum seal on packaging container.	Product has been distributed at retail level.	Withdraw product. (Due to low pH, product will not support pathogen growth, hence withdrawal only.)
party Brie with sundried tomatoes and pine nuts	Ingredient listing does not mention nuts.	Product has been distributed at retail level.	Consumer level: recall product. (Food may pose a safety risk to allergenic consumers.)
all product types	Complaint of foreign object (e.g., metal contamination).	Product has been distributed at retail level.	Consumer level: recall product. (Food may pose a safety risk.)
all product types	Underweight packaging.	Product has been distributed at retail level.	Withdraw product. (Trade practices breach. Food does not pose a safety risk.)

ABC Farms Recall Policy and Procedure

The purpose of a product recall is to do the following:

- Promote customer and consumer safety.

- Locate and remove unacceptable or questionable products from the market at minimum cost and inconvenience to the customer.

- Comply with applicable laws and government regulations.

- Provide accurate information to customers when appropriate.

Recall classifications:

- **Class I recall**: This is a health-hazard situation where there is a reasonable probability that the use of the product will cause serious, adverse health consequences or death. Priority must be given to the immediate and complete removal of all suspected product from the marketplace.

- **Class II recall:** This is a health-hazard situation where there is a reasonable probability of adverse health consequences from the use of the product but it is not life threatening.

- **Class III recall:** This is a situation where the use of the product will not cause adverse health consequences.

  » **Market withdrawal:** This is the removal or correction by a company's own volition of a distributed product that involves a minor infraction that would not warrant legal action and is not a health hazard.

  » **Stock recovery:** This is the removal or correction of product that has not been marketed or that has not left the direct control of the firm. These products have been determined to be mislabeled, poorly packaged, damaged in shipment, or not up to the standards expected by the customer.

Depth of recalls

- Consumer level: includes household consumers as well as all other levels of distribution.

- Retail level: includes all retail sales of the recalled product.

- User level: includes hotels, restaurants, and other foodservice institutional consignees.

- Wholesale level: refers to the distribution level between the manufacturer and the retailer. This level may not be encountered in every recall situation.

Recall communication

ABC Farms is responsible for promptly notifying each of its accounts that may be affected by any recall. In the event of a recall, this communication will be accomplished by telephone, fax, email, or special delivery letters. These guidelines will be followed:

- ABC Farms will clearly identify the product being recalled. Product, size, lot numbers, codes, vat numbers, and sell-by dates will be provided to enable accurate and immediate identification of the product.

- ABC Farms will explain concisely the reason for the recall and identify any possible hazards associated with the product.

- ABC Farms will provide specific instructions on what should be done with respect to the recalled product.

- ABC Farms will provide a ready means for the recipient of the communication to report to the recalling firm whether it has any product.

- When necessary, follow-up communications will be sent to those who fail to respond to the initial recall communication.

Public notification

The purpose of public notification is to inform the public that a product is being recalled. If the need arises, public notification will be handled by press release and through the general news media.

Effectiveness checks

The purpose of the effectiveness checks is to verify that all parties that have been alerted about the recall have received notification about the recall and have taken appropriate action. Affected parties will be contacted by personal visits, telephone calls, letters, faxes, or a combination thereof. Contact coverage may vary based on the type of recall. The levels are listed below.

- Level A: All, or 100 percent, of the known recipients of the product are to be contacted. Level A effectiveness checks will be warranted in Class I recalls.

- Level B: Some percentage of the total number of product recipients will be contacted on a case-by-case basis. Level B effectiveness checks will be warranted in Class II and Class III recalls.

- Level C: No effectiveness checks will be conducted when this level is designated. This level is designated for Class III market withdrawal or stock recovery recalls.

Termination of recall

This is when the product subject to the recall has been removed and proper disposition or correction has been made.

Recall responsibilities

The decision to initiate a recall is the responsibility of the President of ABC Farms, Joe Smith, and/or the Vice President, Jill Jones. The person initiating the recall will designate a recall coordinator who is responsible for forming a recall team and assigning the various tasks that must be performed.

If you are a small operation, this job will fall to you; if you have hired staff, you may want to assign your quality-assurance (QA) manager or sales manager to coordinate the recall since they are the people familiar with inventory control and marketing channels.

The recall responsibilities will be carried out in the following manner:

- The recall coordinator and the recall team will evaluate the facts, information, reports, and test results. From this information, the recall coordinator will determine the necessity, extent, and depth of the recall.

- The recall coordinator will direct the proper personnel to proceed with the notification of affected customers, regulatory agencies, and ABC Farms staff regarding the recall of product and what procedures will be followed. The recall coordinator will directly supervise, document, and confirm the return and disposition of all recalled product.

- The recall coordinator will implement the effectiveness checks required and will also prepare and send out the proper recall letters and press releases for the level of recall initiated.

- The recall coordinator will maintain a log of events, evaluate the progress of product recall, and identify and implement procedures for terminating the recall. The final decision as to when the recall has been satisfied will rest with the recall coordinator.

- The recall coordinator is responsible for direct communication with insurance carriers and legal counsel that may be necessary as a function of the recall. He or she is responsible for notifying the proper authorities and agents and ensuring that all legal requirements are met to satisfy the recall.

RECALL PROCEDURES

1. **Receive complaint**: All customer complaints are directed to either the President or Vice President of ABC Farms. All complaints are evaluated for their severity and a decision is made as to disposition. If the complaint is investigated fully and is thought to be of a potentially serious nature, a formal recall procedure may be implemented.

2. **Assess public health significance**: Based on the source of the complaint, the evidence received, and tests performed, the recall coordinator will immediately determine the level of recall and proceed with the appropriate procedures.

3. **Formally notify regulatory agency**: The recall coordinator works closely with regulatory agency personnel. In the event of a recall requiring communication with a regulatory agency, the recall coordinator will be responsible for such notifications as well as notifying the appropriate legal counsel and insurance carriers. The notice to regulatory agencies will include the following information:

   › complete and accurate product identity

   › reason for the recall and details about when and how any defect or deficiency was discovered

   › evaluation of the risk associated with consumption of the product and the method of making the evaluation

   › total amount of product produced and during what time frame

   › estimate of how much of the product is in distribution and how long it has been in distribution

   › area of the geographical distribution of the recalled product by state and, if exported, by country

   › names of distributors and customers that received the product

   › copies of any company correspondence with distributors, brokers, or customers relating to the recall strategy or actions, and a copy of any proposed press release

   › name, title, and telephone number of the recall coordinator for the company

4. **Determine classification and depth of recall**: After a formal recall has been initiated, the recall coordinator will classify the recall and form a recall team. The classification of the recall will dictate what events and procedures will occur next. If the classification is a Class I, the recall

coordinator will issue the appropriate formal public notices and follow the appropriate protocol for such a recall. If the recall is of a lesser class than I, the recall coordinator will direct the appropriate personnel to start the removal of the product from the marketplace. The recall coordinator will supervise the disposition of all recalled or withdrawn product.

5. **Monitor recalls and effectiveness checks:** The recall coordinator is responsible for monitoring all recalled product and ensuring that all suspect product is recovered. He or she is also responsible for making the appropriate consumer,

retail, and distributor credits for such returns. The recall coordinator will direct company personnel to follow up with affected customers. The recall coordinator will document the extent of the recall, notify all appropriate individuals as to the final disposition of recalled product, and notify all as to the outcome of the recall or product withdrawal.

Figures 6.6, 6.7, and 6.8 provide a sample customer notification log, product identification sheet, and recall notification letter respectively. Figure 6.9 indicates who might be a part of a recall team.

Figure 6.6. Customer notification log.

Customer Notification Log

Name of customer: _____

Date called: _____ Time called: _____

Name of person contacted: _____

Date info faxed: _____ Time faxed: _____

Amount in their possession: _____ (pounds, cases, pack size, etc.)

Amount sold to their customers: _____ (pounds, cases, pack size, etc.)

Can this be recovered from the marketplace? _____ (yes, no)

If yes, who has it and how much? _____

_____

If contact is not established on initial call, please follow up.

Figure 6.7. Recall product identification sheet.

Recall Product Identification Sheet

Date: _____   Brand name: _____

Product name: _____   Package (type and size): _____

Package code (use by/sell by): _____   Packaging date: _____

Case code: _____   Count/case: _____

Production date: _____   Amount produced (lbs/cases): _____

Amount held at establishment: _____   Distribution level (institutional, retail, etc.): _____

Distribution area: _____

Figure 6.8. Sample recall notification letter.

DATE:

Customer firm name and address

ATTN: *Contact person name and title*

Re: RECALL OF *type of product*

Dear Sir or Madam:

This letter is to confirm, per our telephone conversation, that *ABC Farms* is recalling the following product(s) because *specify recall reason*:

(Describe the product, including name, brand, code, package size and type, etc.)

We request that you review your inventory records and then segregate and hold the above product(s). If you have shipped any of this product, we request that you contact your customers and ask them to retrieve the product and return it to you. Once you have retrieved all of the product, please contact us. We will arrange to have the product shipped to our facility. Please **do not** destroy the product. We will credit your account for product returned.

Your prompt action will greatly assist ABC Farms. If you have any questions, please do not hesitate to contact our *company recall coordinator, name,* at *phone number*.

Thank you for your cooperation.

Sincerely,

Joe Smith, President

Jill Jones, Vice President

ABC Farms

Food-Safety Strategy No. 11: Establish Process Control Protocols

Establish process control protocols and sample every lot or batch of cheese for pH, salt, and moisture.

Sample every lot or batch of cheese at the appropriate time to be sure you are reaching the pH, salt, and moisture content described in your make process. These three parameters are critical for inhibiting undesirable microbial growth and reducing the risk of foodborne pathogens surviving in your cheese.

But first, a word about "cheese recipes." Cheesemaking is more than a recipe—it is a process. Cheesemaking is not like baking a cake. It has far more complex and long-term chemical and enzymatic reactions. The matrix of fat, protein, and water in the cheese is being acted upon by bacteria and enzymes from the moment the cheese is made until it is consumed. It should be thought of as a living thing and monitored from start to finish to ensure that you are meeting the benchmarks of your particular make process. This is true for all cheeses. Monitoring is critical to food quality and food safety.

As mentioned in the introduction, this is not a book about food technology, but it is about shaping a successful business. Unless you keep track of these key parameters in your cheesemaking process, you won't be able to account for either your success or failure, and you won't be able to control product quality or safety. So we will spend some time here talking about monitoring the cheesemaking process.

Cheese process controls

There are three compositional properties of cheese that not only influence ripening but also affect microbial growth. These compositional properties are salt, moisture content, and pH (acidity). Measuring these properties to be sure the results are consistent is called process control. This will not be an in-depth discussion of food chemistry, but it will cover the importance of good process controls as a means of assuring food safety.

Charting pH

As you outline your cheese make process, you should be able to chart the pH of your cheese from start to finish. This is especially critical in the vat. You will establish what is "normal" acid development for your make process by charting pH over about 50 vats of cheese. By accumulating information about what the pH looks like over time when your process works correctly, it will help you see immediately when there is a problem. Maybe one time you will have an old starter, or there will be additional water that got into the milk line, and these can affect how much acid develops and how long it takes to develop. You want to know what is normal for your cheese for several important reasons.

Low pH will lead to lower microbial numbers and less diverse microbial populations. This happens, in part, when an increase in curd acidity (lowering the pH) causes the protein matrix to shrink, expelling whey. The whey holds dissolved lactose, and the lactose is a key nutrient for microbial growth. The acidity influences enzymatic action, which is important to the type of molds or yeasts you are working with. For example, a pH decrease from 6.0 to 5.7 reduces smear formation on washed rind cheeses from *Brevibacterium linens,* and a pH increase from 4.6 to 4.8 strongly favors yeast and mold formation from *Geotrichium candidum, Penicillium roqueforti,* and *P. camemberti.*

Tracking pH through your make process will assure that you have created conditions for the microbial activity you want, and if the process gets off track, pH can alert you to problems you may encounter later on with spoilage or other quality problems. Because pH influences moisture, it is linked with salt and water activity (a_w).

Measuring salt concentration

Proper salt concentration is important for a uniform cheese flavor, and it influences the elasticity of cheese (think Mozzarella) depending on pH. Salt also provides a barrier to the growth of pathogens and certain spoilage microorganisms. Salt concentration is usually determined several days after manufacture to permit time for equilibration. It takes time for the salt to become evenly distributed throughout the cheese. In the case of brine- or surface-ripened cheese, uniform salt distribution may never be achieved. For Cheddar cheese and other vat-salted cheese, representative samples for accurate determination of salt content can usually be obtained as early as 7 days after manufacture. Insufficient salt can lead to higher microbial numbers and more diverse microbial populations, and it can lead to higher "free moisture" because water molecules are not bound in the cheese.

Measuring water activity

The higher the moisture content of cheese, the more retention of residual lactose there is. Lactose is a perfect microbial food and so can lead to higher microbial numbers. With high moisture, there is a greater risk of abnormal fermentations from nonstarter bacteria. An important measure of moisture content is water activity (a_w). This measures the amount of liquid water available for microbial growth or chemical reactions in a food or solution. For example, pure water has an a_w of 1.00.

Water activity is reduced by dissolved substances like salt, and it varies directly with the number of dissolved molecules in the water rather than the molecular weight of a substance. Both large molecules (proteins) and small molecules (sugar and salt) affect water activity. An example of a food with low water activity is jam. Jams are preserved by their high sugar content, yielding low a_w. You should know what a_w you want in your final product, and monitor it. It will go hand in hand with both salt and pH levels.

Controlling moisture, salt, and pH is essential for managing cheese ripening and food safety. Each influences the other, so you must control all three. Good process control will minimize process defects and help you give your customer a consistently wonderful product. As you begin your product development, be sure you develop product specifications for a_w, salt, and pH—and test each lot against your standards right from the start of your cheese production. When business gets going great guns, don't be tempted to push your cheese output beyond your production and process controls capacity. Paul Kindstedt of the University of Vermont has covered in depth the technical aspects of measuring these quality parameters in his book *American Farmstead Cheese* (Kindstedt 2005).

Tables 6.4 and 6.5 show how the a_w for microbial growth and the normal a_w for a variety of cheeses are closely related. This means that producing a safe cheese relies on multiple factors, including a_w, pH, salt, and sanitary manufacturing practices to keep undesirable bacterial levels low.

Tables 6.6 and 6.7 show the range of pH, water, and salt content of the same cheese varieties made by different manufacturers. You can see in the table for Cotija cheese that the moisture varies by almost 13 percentage points. For the Chèvre, salt varies by as much as 400 percent and moisture by more than 200 percent. Although some variation is expected with farmstead cheeses, especially when grazing conditions change, salt content is controlled by humans in the cheesemaking process.

"The most unexpected thing that ever happened to me was my one milk producer giving 30 days' notice. We had to buy the herd, lease their dairy, and run two dairies for a while!"

Cheesemaker Story

Food-Safety Strategy No. 12: Obtain Product Liability Insurance

Food product liability insurance
If you follow the first 11 steps with enthusiasm, you may never need the protection of step 12. But regardless of the pride and care you take in your cheese manufacturing, you should still protect yourself against financial loss.

Definition
Product liability insurance means protection against financial loss arising out of the legal liability incurred because of injury or damage resulting from the use of a covered product.

If you plan to sell through a certified farmers' market, a grocery distributor, or specialty retailer, you will be asked to have food product liability insurance as a condition of doing business. If you market your product directly from your farm, you may decide not to carry liability insurance. However, if your product causes injury to someone and they take legal action against you and you don't have insurance, you are exposing your business and family to financial risk that could ruin you.

Most food product liability policies are written for a minimum of $1 million per occurrence and $2 million aggregate liability for the term of the policy (usually 1 year). The cost of the policy is based on the kind of product manufactured (dairy products), process controls (like third-party HACCP certification), prior claims, estimated gross sales of the business, and the number of years in operation.

Because each food processor is unique, there is no standard rate for "cheese manufacturers." To determine your costs, you need to work with an insurance carrier to complete an application and submit it for analysis. Depending on the size of your business and other factors, the insurance could run into tens of thousands of dollars.

We have discussed product liability insurance because not everyone may be familiar with it. This type of coverage will be in addition to general liability insurance, workers' compensation, automobile insurance, and property insurance.

Table 6.4. Minimum water activity for microbial growth

Type of organism	Water activity (a_w)*
most bacteria	0.85–0.91
most yeast	0.87–0.94
most molds	0.70–0.80

Note: *a_w measures the amount of liquid water available for microbial growth or chemical reactions in a food or solution. For example, pure water has an a_w of 1.00. Water activity is reduced by dissolved substances like salt, and it varies directly with the number of dissolved molecules rather than the molecular weight of a substance.

Table 6.5. Water activity for select cheeses

Cheese type	Water activity (a_w)
Brie	0.980
Munster	0.977
Saint-Paulin	0.968
Edam	0.960
Cheddar (mild)	0.950
Parmesan	0.917

Table 6.6. Moisture, pH, and salt profiles of Cotija cheese made by different manufacturers

Water %	pH	Salt %
37.60	5.74	2.51
32.93	5.54	2.89
45.79	6.25	3.82
39.10	4.86	2.12
34.90	5.76	2.51
36.30	5.74	3.16

Table 6.7. Moisture, pH, and salt profiles of Chèvre made by different manufacturers

Water %	pH	Salt %
50.30	5.08	2.01
26.12	5.64	1.74
66.00	4.00	0.78
56.28	3.79	0.82
67.60	4.01	0.50
52.50	4.78	1.21
30.90	4.84	1.58

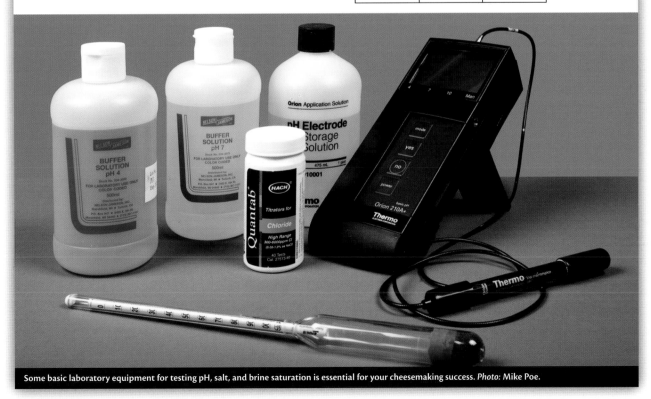

Some basic laboratory equipment for testing pH, salt, and brine saturation is essential for your cheesemaking success. *Photo:* Mike Poe.

MANAGEMENT AND MARKETPLACE RISK

Now that we have covered food-safety and liability issues, there are still other elements of risk to discuss. These are product and market diversity and the stability or security of your physical plant.

What follows here is a story that could happen to anyone. In 2000, some new artisan cheese-makers in the Rocky Mountains sold their first cheeses. Between 2001 and 2005, all of their cheeses won industry awards (including the World Cheese Awards in London). They received lots of press in publications and media such as the *New York Times, Real Simple, The Food Channel*, MSN, and many other outlets. Business was booming, and they could hardly keep up with orders. By 2006, they were out of business. So what happened?

In 2005, the company lost a customer that represented around 50 percent of annual revenues. Attempts were made to replace the customer, but while several new accounts were landed, none approached the volume of the lost customer. As a result, they experienced several consecutive months of negative cash flow.

A large specialty food store chain was the client that represented 50 percent of annual revenues. According to the company's cheesemaker,

this chain decided to stop carrying the products when the cheesemaker could not keep up with demand. When orders and revenues decreased by 50 percent, the cheesemakers could not cover the expenses they had encumbered to ramp-up cheese volume (to meet orders from this food store chain). At the same time, the building they rented for their production facility was slated for demolition by its owner. According to the company's press releases, they also attributed cash flow problems to cost increases for labor, milk, fuel, and shipping. Real estate prices in northern Colorado (whether renting another building or buying land and building their own plant) also put them at a competitive disadvantage.

Let's talk about how these cheesemakers might have been able to minimize their risk.

Long-Term Leases and Food Processing

When looking for a facility (if you don't build it or buy it), you will need to negotiate a long-term lease that includes a renewal option and some rent control language to protect you against inflation if you exercise your lease-renewal option. If a landlord can't offer you these items, then you know you are accepting a high level of risk to locate your plant there. In food processing, 5-year lease

terms with renewal options would be considered a bare minimum. There is a significant investment required (see plant layout in Chapter Four) to take a warehouse space and bring it up to code to be a dairy processing plant. In addition to the infrastructure costs, the equipment design and layout will be specific to the facility.

If someone offers you a price (per square foot) that appears to be unrealistically low when you check out comparable real estate, you should ask yourself why. Complete your market research and cash flow projections to see what your facility costs will be as a percentage of overhead costs. Talk to other cheesemakers and find out what this cost is for them. No matter what the cost, you have to sell enough cheese to cover the expense—and you can't move your plant every 5 years.

Concentrating Risk Through Too Few Markets

Anyone would be excited to get a big contract with large food store chains like Trader Joe's or Costco. It is a sign that you have really arrived as a business. But are you ready for all that "success"? As we discussed in the chapter on marketing, there are some downsides to that success. Large contracts with Trader Joe's or Costco may require you to commit to a specific volume for order fulfillment, and if you can't keep up, there is somebody ready to take your place. In addition, their markups are small, so they will negotiate for a very low wholesale price. You will need to sell a large volume at a discount in order to make the same net revenue you would with a smaller sales volume at a higher per-pound price. In the case described above, committing more than 50 percent of the company's business to a single vendor did two things. The investments that the company made for expansion were not justified by the rest of their cheese sales volume or customer base, and the added debt taken on to do the expansion left the company extremely exposed to risk if they lost the one customer.

Plan for Growth, and Then Manage It

As mentioned in the introduction, most cheesemakers experience double-digit annual growth as soon as they launch their business. That is great, but the best way to grow is to have a variety of outlets for your products and a variety of products. When researching the cheese you want to make, think about how minor modifications in the make process will create a variety of cheeses to sell. They may be differentiated by age, you may sell some fresh and some aged cheeses, or you may make cheeses flavored with herbs and garlic. However you do it, just remember not to have all your cheese wheels in one basket. If you plan to focus on direct sales, sell at several farmers' markets, not just one—and if you can, look for markets that are in a variety of regions with slightly different clienteles.

Contingency Plans

One of the worst things that can happen to a firm is to experience something unexpected and not know what to do about it. Therefore, it is necessary to question every aspect of your plan. You should come up with a series of "what if" questions. What if sales do not come up to expectations? What if the demand is so overwhelming that you can't keep up with supplying the product? What if competitors react to your entry into the market by lowering prices or other competitive reactions?

Contingency plans should be based on all of the things that you planned to do, and actions should be based on monitoring those plans. Not only do you need to know how your sales are going or how your market share is growing, but also you need to know exactly what you will do if you do not achieve your goals.

The best contingency plans are the ones that actually state concrete decisions of action, based on monitoring, before something gets out of control.

That means you must identify trigger points. If profits are 30 percent below what you projected them to be after 3 months, then there should be a plan that is put into operation immediately.

Contingency Plan Checklist

- What environmental, technical, or competitive occurrences and disasters could significantly disrupt the organization's operations, facilities, and ability to serve customers? What specific functions, systems, and locations could be affected?

- Which internal decision makers should be notified in the event of an emergency? Who will initiate this notification?

- What advance preparations can the organization make to minimize the effects of an emergency? Who will be responsible for making these preparations? What resources will be needed?

- What steps should the organization take—and in what order—to restore normalcy if an emergency occurs? Who will be responsible?

- What steps should the organization take—and in what order—to continue serving customers and to keep operations going in an emergency? Who will be responsible?

- How and when should the organization communicate with customers, investors, employees, the media, and other stakeholders about the emergency? Who will be responsible?

Like the Scouts say, be prepared.

POINTS TO REMEMBER

➡ While risk and thus liability cannot be eliminated, they can be reduced and managed via a risk-management plan.

➡ Creating a risk-management plan is an essential part of running a successful business.

» Using good manufacturing practices and developing a HACCP plan are critical components of your risk-management plan.

➡ Your risk-management plan involves numerous strategies to reduce risk to your product, your employees, and your customers, thereby reducing your liability.

➡ You must rely on your management team— particularly your lawyer and insurance agent—to help determine your liability needs.

➡ There are laws and many guidelines to help you take good care of your animals if you operate a commercial dairy.

REFERENCES

Kindstedt, P. 2005. American farmstead cheese: The complete guide to making and selling artisan cheeses. White River Junction, VT: Chelsea Green Publishing Company.

U.S. Food and Drug Administration (FDA). 2004. GMPs–Section four: Common food safety problems in the U.S. food processing industry: A Delphi study in good manufacturing practices (GMPs) for the 21st century –food processing. FDA Web site, http://www.fda.gov/ Food/ GuidanceCompliance RegulatoryInformation/ CurrentGoodManufacturing PracticesCGMPs/ ucm110914.htm.

USDA Food Safety and Inspection Service (FSIS). 1995. List of Proprietary Substances and Nonfood Compounds Authorized for Use under USDA Inspection and Grading Programs. FSIS Miscellaneous Publication No. 1419.

Chapter Six adapted in part from Williams, John Jr. 2011. Food Allergens: Effectively Managing Your Risks. eSCOPE On-line Technical Bulletin, 17:3. Silliker Food Safety and Quality Solutions Web site, http://www.silliker.com/html/SCOPE/vol17issue3.php.

The section "Food-Safety Strategies" adapted in part from University of California Cooperative Extension Farmstead Cheesemaking Workshops, 2003–2005. Glenn County Cooperative Extension, Orland, CA.

The sections "Food-Safety Strategies," "Sample Daily Sanitation Report," "Sample Master Sanitation Schedule," "Food Recall Plan," and "Process Controls" adapted in part from Food Safety Training for Cheesemakers: Prerequisite for HACCP Workshops, 2004–2005. Glenn County Cooperative Extension, Orland, CA.

The section "Model Sanitation Standard Operating Procedures" adapted in part from The Seafood Network Information Center (Seafood-NIC), a part of the University of California Sea Grant Extension Program.

The section "Points to Remember" excerpted, with permission, from George, H. and E. Rilla. 2011. Agritourism and Nature Tourism in California. 2nd ed. Oakland: University of California Division of Agriculture and Natural Resources, Publication 3484.

CHAPTER

Seven

Regulations

Barbara Reed, Ellie Rilla,
and Holly George

Chapter Goals 122

Regulatory Overview 122

Local Jurisdiction 122

State Jurisdiction 124

Federal Jurisdiction 129

Points to Remember 131

REGULATORY OVERVIEW

Local, state, and federal government agencies will regulate your cheesemaking operation. California examples are used here because of the state's strict standards. Other states have similar agencies and laws governing cheesemaking.

Municipal codes and ordinances describe the procedures for complying with local regulations that apply to general land use, construction, environmental, and business issues. Unlike in the case of many other businesses, local government does not regulate the internal operation of a cheese plant. Specific state and federal laws will have jurisdiction over your cheesemaking operation for sanitary design (including some elements of construction, operation, inspection, and licensing), and they will supersede local control.

To start a cheese-manufacturing operation, you must comply with all relevant local, state, and federal requirements.

The following sections describe the broad categories that apply to dairy processing.

Local Jurisdiction

- land use and development

- environmental health

- public health and safety

- business licenses and taxes

- weights and measures

- farmers' markets

State Jurisdiction

- commercial milk production and harvesting

- milk handling (buying or selling of milk)

- milk sales and raw milk laws

- milk quality

- milk hauling

- bacteriological and antibiotic testing

- pasteurization licensing

- cheese plant design and construction

- cheese plant operation

- product labeling review

- organic registration

- environmental sampling program

Federal Jurisdiction

- organic certification

- standards of identity for dairy products

- nutritional and labeling information

- Pasteurized Milk Ordinance and interstate milk shipments (for Grade A market milk products)

- Public Health Security and Bioterrrorism Preparedness and Response Act of 2002

- good manufacturing practices (GMPs)

- FDA inspections

LOCAL JURISDICTION

Land Use and Development

Each county has a general plan that describes its land use policies. The plan identifies land use zones (agricultural, commercial, and residential, for example). The county zoning ordinance spells out regulations that implement the general plan and describes how land parcels in each land use zone can be used. The codes address rules on existing buildings, new construction, signage, and more. Counties are responsible for incorporating provisions of the California Environmental Quality Act (CEQA) into their county codes. Counties assess and address the environmental impacts of development projects, including water quality, waste disposal, septic systems, and traffic. The zoning ordinance distinguishes between uses that are permitted by right and uses that are subject to conditions, review, and approval.

A cheese plant may be considered an agricultural processing facility, a manufacturing plant, or a manufacturing plant with retail store. This will depend on the county where you build, whether you are a farmstead operation or purchasing milk and not operating a farm, and if you plan on making direct sales from the farm (or plant). You need to talk to your local planning department now and have this conversation long before you start drawing up plans.

Some counties may not allow retail operations on agriculturally zoned property, and they may consider cheesemaking to be manufacturing, not agricultural processing. Your activities may trigger a request for a variance from the zoning code.

If you build a new manufacturing plant (whether on your farm or another location), you will be required to prepare an environmental impact report (EIR). This report will be developed by a third party, based in part on information you provide on the size and scope of your plans. Although you will have to pay for the consultant services, your payment will be given to the city or county in which the facility will be built, and the planning department will be responsible for hiring the consultants to prepare the report. The EIR will address potential environmental impacts, such as increases in traffic and water or air pollution. Unless there are special circumstances, you will have to reduce or eliminate your impacts in order to proceed with your project.

Building Codes

Depending on the desired (new) uses of your property, the zoning code or conditions of the variance may trigger requirements for additional paved areas, building and parking accessibility design, and public restrooms. All public-use structures (like retail stores) must conform to accessibility standards of the Americans with Disabilities Act. These modifications can add thousands of dollars to your project. It is also possible that you could be denied a variance to operate at the location you have selected. If so, you may have to modify your operational plans or find another site. If you can't develop on your own land, the costs of start-up may increase dramatically.

Environmental Health/ Public Health and Safety

The local health department will deal with most common public health concerns related to the sanitation and safety of food processing, such as the water source used in the plant (municipal or well water) and sewerage or septic permitting, fire codes, and design for accessibility.

Each county has its own septic standards and usually requires a permit to repair, upgrade, or construct a septic system. If you do not have a septic system, you will be required to have the site and soil evaluated to determine what type of system can be permitted.

Besides the amount of water your daily plant sanitation program will generate, whey will be the largest waste by-product of your cheese operation. The proper reuse or disposal of whey will be essential for environmental protection. Because cheese plants can have significant wastewater production, your facility might require review and permitting by the Regional Water Quality Control Board. You may be prohibited from using your septic system to handle your whey production. For every pound of whole milk cheese produced, 9 pounds of whey is generated. So if you plan on making 100,000 pounds of cheese per year, you will need to reuse or dispose of almost 1 million pounds of whey. Disposal of whey is an environmental concern, and it has been the subject of a multi-million dollar enforcement action in California by the Regional Water Quality Control Board. Although you won't be producing as much whey as one of the largest cheese plants in the United States, you also won't have the money to invest in microfiltration and water recycling systems. You may need to set up a retention pond for whey and wash water. If you are producing sweet whey, you may be able to feed it to livestock.

If you plan on operating a retail food facility, you will also come under the jurisdiction of the California Uniform Retail Food Facilities Law (CURFFL) or the equivalent in your state. This state law governs food-safety requirements for food handling, equipment, and storage. CURFFL is enforced at the local level by the county environmental health department. At least one person in the enterprise must hold a "food handler's certificate." This person may be either the business owner or an employee. County health departments offer food-safety classes for acquiring a food handler's certificate.

Business Licenses and Taxes

You will need to have a business license and file a fictitious business statement under your enterprise's name. This is necessary if you do business under a name that is not your own or is other than your legally registered business name. Commonly known as a "doing business as" or DBA statement, a fictitious business statement is registered with the county clerk-recorder.

Weights and Measures

Both the county and state have jurisdiction over direct marketing. You will have to license your scales annually with county weights and measures (part of the agricultural commissioner's office). Your scales are inspected to be sure they are calibrated and working properly.

Organic Registration

Federal laws define organic dairy production. According to the California Department of Food and Agriculture (CDFA), "every person engaged in the state of California in the production or handling of raw agricultural products sold as organic, and retailers that are engaged in the production of products sold as organic, and retailers that are engaged in the processing, as defined by the National Organic Program (NOP), of products sold as organic, shall register with the State Organic

Program (through the county agricultural commissioner). If the expected organic gross sales exceed $5,000, certification is required." (See the "Federal Jurisdiction" section below regarding organic certification.)

Farmers' Market Permits

If you operate a booth and sell cheese at a certified farmers' market, you must obtain a Certified Producer's Certificate (CPC) from the county agricultural commissioner in the county where your plant operates.

STATE JURISDICTION

The licensing unit of the Milk and Dairy Food Safety Branch (MDFSB) handles the issuing of licenses and permits pertaining to milk production and processing. This includes all of the dairy farms and processors in California that produce milk products or products resembling milk products. Other states have similar licensing units.

Note that cow's milk is subject to additional regulations compared to goat's and sheep's milk. This is because cow's milk is regulated by state and federal marketing orders. Goat's milk and sheep's milk are priced as per agreement between buyer and seller, and they are not subject to marketing orders. To obtain milk from a particular cow dairy, a buyer may agree to pay a premium for that milk.

Commercial Milk Production and Harvesting: Permits and Inspections

The Grade A or Grade B permit covers your milking facility and milk storage tanks on the dairy. Your permit may be issued by the county environmental health department on behalf of the CDFA. Registered dairy inspectors are responsible for inspecting market milk (Grade A) and manufacturing grade (Grade B) dairy farms for conformance with quality and sanitation requirements. They will take milk samples and test them for bacterial content and the presence of contaminants, and they will routinely collect samples of finished products from retail outlets and analyze them for compliance with standards. Inspectors also investigate consumer complaints and follow up with appropriate actions. They can condemn milk and milk products that may be unfit or unsafe for human consumption.

Milk Quality Standards

We advise new processors to produce and use milk that meets Grade A standards. The quality and consistency of your product will be better when using milk that meets high sanitary and bacteriological standards. Grade A milk is produced under sufficiently sanitary conditions to qualify for fluid (beverage) consumption. (It is also referred to as fluid grade milk). Only Grade A milk is regulated under federal milk marketing orders (cow's milk only). More than 90 percent of all milk produced nationally is Grade A. Therefore, much of the Grade A milk supply is used in manufactured dairy products. Grade B milk (also referred to as manufacturing grade) does not meet fluid grade standards and can only be used in manufactured products.

Milk Handling: Buying or Selling of Milk

A milk handler's license is required of every person who purchases, handles, receives, or otherwise acquires ownership, possession, or control of market milk or manufacturing milk in unprocessed or bulk form for the purpose of manufacturing, processing, sale, or other handling.

Obligations of handlers

The following list highlights some of the obligations of processors/handlers conducting business in California and other states:

- Obtain a milk handler's license prior to purchasing bulk milk.

- Obtain a surety bond prior to purchasing bulk milk from cow's milk producers. This bond entitles you to buy cow's milk. The bond does not apply to goat's or sheep's milk because there is no state or federal marketing order that regulates goat's and sheep's milk or guarantees protection against payment defaults.

- Execute a milk purchase contract with the producer prior to purchasing bulk milk. Pay producers not less than minimum class prices, as established by the CDFA (cow's milk only).

- Pay producers in a timely manner and in accordance with the provisions in the contract. Pay assessments on milk purchases to the Milk Pooling Branch and to marketing programs as charged (cow's milk only).

- Maintain records for 3 years on milk purchases, uses, costs, and other financial transactions with producers and wholesale customers.

- Comply with report filings with Dairy Marketing Branch as required.

Milk Sales and Raw Milk Laws

If you are a licensed producer or handler of milk, take note of specific laws that cover raw milk sales in the western states. Generally, you cannot buy raw milk for commercial cheese production from an unlicensed producer, nor can you sell raw milk to anyone who comes to your farm. Be sure you are in compliance with the laws.

California laws do not allow for any commercial transfer of milk between parties unless they are licensed producers, handlers, or processors. Check to see what your state laws allow.

California Food and Agriculture Code: Division 15, Chapter 13, Article 1, Section 35283 states that it is a felony to knowingly

» (a) process without pasteurization any milk or milk product which is required…to be pasteurized

» (b) manufacture or process for resale any milk or milk product in a…plant which is not licensed…

» (c) provide milk or milk products to any person for the manufacturing or processing for resale of any milk or milk product unless that person is licensed…pursuant to Chapter 12

(To read the passage in full, see the CDFA's California Law Web site, http://www.leginfo. ca.gov/cgi-bin/calawquery?codes ection=fac&codebody.)

Milk Hauling

This involves the regulation of pickup and transportation of milk, as well as the licensing of the truck and of the person driving the truck. A permit is required to transport unpackaged market milk or unpackaged market milk products (bulk milk hauler). A permit is required for each tanker used in the bulk transport of unpackaged market milk or unpackaged market milk products. The permit is only valid for 1 year and must be permanently affixed on the tanker for which it is issued. This permit is specific to the truck, whether it is an 18-wheeler hauling 50,000 pounds or a pickup truck with a 350-gallon refrigerated tank on a flatbed. The driver must also hold a permit and pass an examination to be a milk sampler and hauler. Bulk milk tanker trucks must be cleaned or sanitized at a facility holding a valid bulk milk tanker truck cleaning facility permit, a licensed milk products plant, or a permitted market milk dairy farm.

Note that the handling and transportation costs of cow's milk are regulated by the CDFA. Even if you own your own truck, there will be special costs associated with the permitting and operation of that truck.

Bacteriological and Antibiotic Testing

Milk quality must be assessed by a certified technician. Every person who makes any bacteriological determination of milk or cream must hold a technician's license if the test is to be used as a basis of payment or determining value. For more information about milk testing requirements, see Appendix N of the Pasteurized Milk Ordinance on the FDA Web site, http:// www.fda.gov/Food/ FoodSafety/ Product-SpecificInformation/ MilkSafety/NationalConference onInterstateMilkShipments NCIMSModelDocuments/ PasteurizedMilkOrdinance2007/ default.htm.

Pasteurizer Licensing

Every person who operates equipment that pasteurizes milk must hold a pasteurizer's license. To obtain a pasteurizer's license, an applicant must receive passing grades on written and oral exams and must demonstrate the ability to properly pasteurize milk or its products in a practical examination. A general license covers pasteurization by all approved methods of pasteurization, and a limited license applies only to vat pasteurization. The pasteurizer equipment must be tested and certified by the CDFA.

Cheese Plant Design and Construction

Submit your cheesemaking facility plans to the Milk and Dairy Food Safety Branch-Licensing Unit well ahead of construction so that the dairy inspector can review them. Federal law requires that plans be submitted to the regulatory agency for written approval before construction begins. In California, minimum construction standards for new milk products plants must include a separate room for each of the following operations or equipment:

- receiving and weighing of milk or cream, and washing and sterilizing of containers in which milk or cream is received

- pasteurization, processing, cooling, and manufacturing

- washing and sterilizing bottles or cans that are used in the delivery of milk or cream to the wholesale or retail trade

- storing of supplies

- bacteriological and chemical analyses

- adequate and efficient cold storage room

- boiler, compressor, and other rooms mechanically equipped and operating

- sanitary facilities and changing areas, including toilets, lavatories, and lockers

Cheese Plant Operation

Permits are needed to cover the operations and inspections of your cheesemaking facility. The CDFA licensing unit inspects processing plants and collects samples of milk and milk products to assure consumer safety. They ensure that the tests used to determine the basis of payment for cow's milk are accurate. They also evaluate milk plants and laboratories for the U.S. Food and Drug Administration (FDA) for dairy products in interstate commerce.

Product labeling review (specific to dairy products)

This review will ensure that you are in compliance with state and federal labeling laws. You must have package labels that identify your product, your business name, address, and plant number, as well as the package weight (net weight if the cheese is packed in oil or water). Section 32912.5 of the California Food and Agriculture Code states that "sample copies of all labels to be used in connection with advertising and consumer sales of milk, milk products, frozen desserts, cheeses, and products resembling milk products shall be submitted to the secretary for approval prior to the use of those labels."

If your company is in California and has a label that needs to be reviewed, you can download the "Label Review Checklist" (PDF 153 KB) from the CDFA Web site, and submit the form, billing information, and three color copies of the label to

Dairy Program Coordinator
Department of Food and Agriculture
Milk and Dairy Food Safety Branch
1515 Clay Street, Suite 803
Oakland, CA 94612

Phone: (510) 622-4810

Web site: http://www.cdfa.ca.gov/ahfss/Milk_and_Dairy_Food_Safety/Label_Review.html

For more information on the FDA labeling laws, go to http://www.fda.gov/Food/Labeling Nutrition/FoodLabelingGuidance RegulatoryInformation/default.htm or contact your state Department of Food and Agriculture.

Milk Pricing Classes

These are classification categories for cow's milk in California. (Check your state for classification categories.) There are five classes of milk—and the pool prices for milk are determined in part by how much milk the processors use in any class.

The five classes of milk

Class 1: Milk used in fluid products, including whole, lowfat, extra light, and nonfat milks.

Class 2: Milk used in heavy cream, cottage cheese, yogurt, and condensed products.

Class 3: Milk used in ice cream and other frozen products.

Class 4a: Milk used in butter and dry milk products, such as nonfat dry milk.

Class 4b: Milk used in cheese, other than cottage cheese.

Table 7.1. California state agencies and their responsibilities

State agency	Responsibility
Department of Industrial Relations	Sets occupational health and safety standards. *(Employees must have a health and safety plan.)*
Department of Health Services	Enforces the California Health and Safety Code. Inspects food-processing facilities regarding products exported from the country.
CALTRANS	Reviews development of proposals for traffic-flow impacts. Issues permits for state highway signs.
California Department of Food and Agriculture (CDFA) Milk and Dairy Food Safety Branch	The Milk and Dairy Food Safety Branch (MDFSB) ensures that milk, milk products, and products resembling milk products are safe and wholesome, meet state and federal microbiological and compositional requirements, and are properly labeled. Functions include: providing training and supervision for local Approved Milk Inspection Services to develop statewide uniformity, inspecting dairy farms and milk-processing plants; taking samples of milk and milk products to assure consumer safety; ensuring that tests used to determine the basis of payment for milk are accurate; evaluating dairy farms, milk plants, and laboratories for the U.S. Food and Drug Administration (FDA) for dairy products in interstate commerce; and providing product-grading service for the U.S. Department of Agriculture (USDA). *(Applies to the milk of all dairy species.)*
CDFA Milk and Dairy Food Safety-Licensing Unit	Handles the issuing of licenses and permits pertaining to milk production and processing. This includes all of the dairy farms and processors in California that produce milk products or products resembling milk products. Some of the plant types include frozen, manufacturing (cheese), and soft serve vendors. *(Applies to the milk of any dairy species.)*
CDFA Dairy Marketing Branch	The Dairy Marketing Branch (DMB), working closely with the Milk Pooling Branch, addresses policy issues. The DMB administers the Marketing and Stabilization Plans for market milk in Northern and Southern California (Plans). The branch is organized into five units, each of which concentrates on a specific area of work that contributes to administration of the Plans. The DMB's five units focus on cost of production, manufacturing cost, enforcement, statistics, and economics. *(Applies only to cow's milk.)*
CDFA Milk Pooling Branch	Administers the California Pooling Plan for (cow's) Market Milk and works closely with the Dairy Marketing Branch to promote marketing of (cow's) farm milk. Statewide pooling of revenue from dairy processors and redistribution of those pooled revenues to dairy (cattle) farmers. Helps ensure that dairy farmers receive an equitable and fair price for the milk they produce. The Milk Pooling Branch is divided into three principal units: the Operations Unit, the Pooling Audit Unit, and the Producer Payments Unit.
Regional Water Quality Control Board	There are nine Regional Water Quality Control Boards (Regional Boards). The mission of the Regional Boards is to develop and enforce water-quality plans that protect the state's waters, recognizing local differences in climate, topography, geology, and hydrology. Regional Boards develop "basin plans" for their hydrologic areas, issue waste discharge requirements, take enforcement action against violators, and monitor water quality. Regional Boards are responsible for state enforcement of the Federal Clean Water Act and California state laws relating to water protection. Farmstead operators must obtain an extra use permit when discharging cheesemaking wastewaters to a new or existing wastewater pond.*

Note: *See figure 7.1 note.

Table 7.1 summarizes the state agencies that have jurisdiction over dairy farms and cheese plants.

FEDERAL JURISDICTION

Organic Certification

This requires independent, third-party auditing. To obtain organic certification, a producer or handler must submit an application for certification to an accredited certifying agent. The application must contain descriptive information about the applicant's business, an organic production and handling system plan, information concerning any previous business applications for certification, and any other information necessary to determine compliance with the federal organic food laws. This is usually handled by your state inspector for the FDA. Your inspector should sign off on the label before it is printed.

For complete information about the National Organic Program, go to the USDA's Agricultural Marketing Service Web site at http://www.ams.usda.gov/NOP/indexIE.htm.

Standards of Identity for Dairy Products

The federal government has a set of definitions for cheeses that describe allowable variations in the moisture, solids, and fat content of those cheeses. These definitions can be found in the U.S. Code of Federal Regulations Title 21 (Food and Drugs), Article 133 (Cheeses and Cheese Related Products), viewable on the U.S. Government Printing Office Web site (GPO Access), http://www.access.gpo.gov/nara/cfr/waisidx_04/21cfr133_04.html.

Nutritional and Labeling Information

The same section of the federal law above also covers cheese labeling. This is handled by the state of California for the FDA.

Pasteurized Milk Ordinance (PMO) and Interstate Milk Shipments

The PMO outlines minimum standards and requirements for Grade A milk production and processing, published by the FDA. Grade A standards are recommended by the National Conference on Interstate Milk Shipments (NCIMS). The NCIMS is made up of representatives from state and local regulatory agencies, the dairy industry, and the FDA. As a general rule, the FDA accepts the NCIMS recommendations and incorporates them into revisions of the PMO. The state regulator adopts the PMO standards as a minimum and in many cases requires more stringent standards. The CDFA enforces the PMO in California. Information on the PMO can be found on the FDA's Food Safety Web site, http://www.fda.gov/Food/FoodSafety/Product-SpecificInformation /MilkSafety/default.htm.

Public Health Security and Bioterrorism Preparedness and Response Act of 2002

In the wake of the terrorist attacks on U.S. soil in 2001, the federal government initiated a program to protect public safety and the U.S. food supply. The resulting bioterrorism act requires domestic and foreign facilities that manufacture, process, pack, or hold food for human or animal consumption in the United States to register with the FDA. You must register your facility even if your cheese will never leave California to enter interstate commerce. More information about the bioterrorism act can be found on the U.S. Customs and Border Protection (CBP) Web site, http://www.cbp.gov/xp/cgov/trade/trade_programs/is_initiatives/bioterrorism/bioterrorism_act.xml.

Food Safety Modernization Act of 2011

In January of 2011, President Obama signed into law the Food Safety Modernization Act. Under the new law, food companies will be required to do the following: consider and identify all potential food-safety hazards associated with their products and operations; develop written

Figure 7.1. Regulatory agency worksheet.

Jurisdiction	Agency	Permit or fee	Start-up cost	Annual cost
county	Agricultural Commissioner	scale certification		
		organic inspection		
		labeling inspection		
	Treasurer/Tax Collector	business tax and license		
	Building Department	building permit		
	Planning Department	use permit		
	Environmental Health Department	food-safety permit		
		septic permit		
		well permit		
	Public Works	permit for entry onto county roads		
		signs permit		
	Clerk/Recorder	fictitious name statement		
	Fire Agency	safety and occupancy inspections		
state	CDFA - Milk and Dairy Foods	Grade A dairy permit		
		cheese plant permit		
		milk handler license		
		milk handler surety bond (cow's milk only)		
		milk hauler/sampler permit		
		pasteurizer license and inspection		
		pasteurizer operator examination and license		
		dairy product label review		
		dairy inspections		
		cheese plant inspections		
	California Franchise Tax Board	income taxes		
	California State Board of Equalization	sales and use tax registration		
	California State Department of Industrial Relations	California Occupational Safety and Health Administration: employee safety plan		
	California Department of Transportation	entry onto state highway		
		signage license and fees		
federal	United States Department of Agriculture	organic certification, via a third-party audit		
	Food and Drug Administration (FDA)	Pasteurized Milk Ordinance (accomplished through CDFA Milk and Dairy Foods)	n/a	
		standards of identity for dairy products (accomplished through CDFA Milk and Dairy Foods)	n/a	
		nutrition and labeling information (accomplished through CDFA Milk and Dairy Foods)	n/a	
		Public Security and Bioterrorism Preparedness Response Act of 2002–facility registration	no cost	
		cheese plant inspection		

Note: *The Dairy Practices Council produces peer-reviewed publications related to wastewater management for dairy processing, as well as many other publications related to milk production and sanitation. For a small fee, you can order publications from the Dairy Practices Council Web site, http://www.dairypc.org/guidelinelisting.htm.

plans to respond to each of those hazards; and closely follow those plans to reduce or eliminate such hazards to the greatest extent possible.

Good Manufacturing Practices (GMPs)

GMPs are covered in depth in Chapter Six. They are fully described in the Code of Federal Regulations Title 21, Chapter 1, Part 111, found on the U.S. Government Printing Office's Web site (GPO Access), http://www.access.gpo.gov/nara/cfr/waisidx_10/21cfr111_10.html.

FDA Inspections

The FDA conducts facility inspections independently of any state inspection. Funding and staffing issues with the FDA prevent them from inspecting every food-manufacturing facility every year. However, if a consumer complaint is lodged with your local environmental health department or the CDFA, the FDA may inspect your facility and issue an advisory or enforcement action.

Figure 7.1 is a worksheet that provides a place for you to record inspection and permit costs.

POINTS TO REMEMBER

➡ Numerous regulations—many of them complex—face individuals and families interested in establishing a farmstead or artisan cheese enterprise.

➡ Regulations are part of doing business, and your compliance with them helps protect operators as well as consumers from potential liabilities.

➡ You can best address the regulatory bureaucracy by taking it one step at a time.

➡ Agency Web sites and staff can answer questions, provide information, and help you meet requirements.

➡ A good working relationship with all agency staff is vital, both during the permit application process and during later inspections.

➡ The time required for the permit approval process varies with each operation; therefore, allow for a lengthy procedure.

➡ Regulations are important to the development and cost estimates of a business plan, so identify them early in your planning.

Chapter Seven adapted in part from University of California Cooperative Extension Farmstead Cheesemaking Workshops, 2003–2005. Glenn County Cooperative Extension, Orland, CA.

The section "Points to Remember" excerpted, with permission, from George, H. and E. Rilla. 2011. Agritourism and Nature Tourism in California. 2nd ed. Oakland: University of California Division of Agriculture and Natural Resources, Publication 3484.

GLOSSARY

ACS (American Cheese Society). ACS provides American cheesemakers with educational resources and networking opportunities, and it encourages the highest standards of cheesemaking. Membership in the American Cheese Society is available to anyone involved in the trade or simply passionate about American-made specialty and artisanal cheeses.

ADA (Americans with Disabilities Act). Federal law that ensures that people with disabilities have access to goods, services, and facilities. Under Title III (Public Accommodations and Commercial Facilities), no individual may be discriminated against on the basis of disability with regards to the full and equal enjoyment of the goods, services, facilities, or accommodations of any place of public accommodation by any person who owns, leases (or leases to), or operates a place of public accommodation. "Public accommodations" include most places of lodging, recreation, transportation, education, and dining, along with stores, care providers, and places of public displays.

allergens. Foods or food ingredients that trigger allergic reactions. They are milk, egg, peanut, tree nuts, fish, shellfish, soy, and wheat.

allowances. Discounts given to retailers in exchange for either favorable placement of a product in their stores or the initial or continued stocking of a product.

antibiotic. A substance that kills or inhibits the growth of bacteria.

anti-siphoning device. Prevents the flow of contaminated water into the potable water system in the event of a reverse of water pressure in the system.

approved carrier. Retailer-specified trucking company that meets sanitary, refrigeration, and other requirements.

approved vendor. In order to become an approved vendor, a food manufacturer must meet requirements of the grocery wholesaler or retailer, which may include verification inspections, licenses, or other requirements such as HACCP and liability insurance.

artisan cheese. Cheese made by hand, using the traditional craftsmanship of skilled cheesemakers.

back-flow prevention. See anti-siphoning device.

backhaul. Trucks make multiple pickups and deliveries on the same round-trip in order to make use of space in the truck on the return route.

balance sheet. Statement of the financial position of a business on a specified date.

batch pasteurization. The heating of every particle of milk or milk product to a specific temperature for a specified period of time without allowing recontamination of that milk or milk product during the heat treatment process. This is done using a vat pasteurizer, which consists of a jacketed vat surrounded by either circulating water, steam, or heating coils.

break-even point. Point in time (or in number of units sold) when revenue equals the total cost, and profit begins to accumulate. This is the point at which a business, product, or project becomes financially viable.

brine tank. Tank containing water saturated with or containing large amounts of a salt, especially sodium chloride.

brining. Soaking cheeses in a saturated salt solution.

broker. Person who serves as an agent or intermediary in negotiations or transactions.

BTU (British thermal unit). The amount of energy needed to heat 1 pound (0.454 kg) of water from 39° to 40°F (3.9° to 4.4°C).

capital. Financial capital refers to money used by entrepreneurs and businesses to buy what they need to make their products or provide their services. It also refers to that sector of the economy based on its operation, that is, retail, corporate, investment banking, etc.

case cards. Point-of-purchase display information that includes descriptions of the product attributes and origin.

cash flow. The movement of cash into or out of a business. In accounting, cash flow is the difference between the amount of cash available at the beginning of a period (opening balance) and the amount at the end of that period (closing balance).

CDC (Centers for Disease Control and Prevention). The CDC is one of the major operating components of the Department of Health and Human Services. The CDC's mission is to collaborate in creating the expertise, information, and tools that people and communities need to protect their health. This is done through health promotion; prevention of disease, injury, and disability; and preparedness for new health threats.

charge back. The return of funds to a consumer, which occurs when suppliers sell a product at a higher price to the distributor than the price they have set with the end user.

CIP (clean in place). A method of cleaning the interior surfaces of pipes, vessels, process equipment, filters, and associated fittings without disassembly.

coliforms. Coliform bacteria are commonly used as a bacterial indicator of sanitary quality of foods and water. They are defined as rod-shaped, gram-negative, nonspore-forming bacteria, which can ferment lactose with the production of acid and gas when incubated at 35° to 37°C. While coliforms are themselves not normally causes of serious illness, they are easy to culture, and their presence is used to indicate that other pathogenic organisms of fecal origin may be present.

collateral. In lending agreements, collateral is a borrower's pledge of specific property to a lender, to secure repayment of a loan.

commodity cheese. Also called "government cheese," this processed cheese was provided to welfare and food stamp recipients in the United States during the 1980s and early 1990s. Like all American processed cheese, it consists of a variety of cheese types blended together with other ingredients such as emulsifiers, and may be made in part from Cheddar cheese, Colby cheese, cheese curd, or granular cheese. The cheese often comes from food surpluses stockpiled by the government as part of milk price supports.

contingency plan. An alternative plan devised to anticipate a problem.

cooperative. A cooperative is a business organization owned and operated by a group of individuals for their mutual benefits, or it is a business owned and controlled equally by the people who use its services or by the people who work there.

COP (clean out of place). Equipment that can be removed or disassembled for cleaning.

co-packing. A contract packer, or co-packer, is a company that manufactures and packages foods or other products for their clients. To market and distribute, a co-packer works under contract with the hiring company to manufacture food as though the products were manufactured directly by the hiring company.

corrective actions. A change implemented to address a weakness identified in a processing system.

CPC (Certified Producer's Certificate). A CPC gives you the right to sell fresh fruits, nuts, vegetables, shell eggs, honey, flowers, and nursery stock directly to consumers at certified farmers' markets without the usual size, standard pack, container, and labeling requirements.

credit terms. Standard or negotiated terms (offered by a seller to a buyer) that control 1) the monthly and total credit amount, 2) the maximum time allowed for repayment, 3) the discount for cash or early payment, and 4) the amount or rate of late payment penalty.

cross-contamination. A leading cause of foodborne illness, this is the transfer of harmful bacteria or other contaminants to food when proper food-safety procedures are not followed.

cross-docking. Distribution method in which the goods flow from receiving to shipping, thus eliminating storage. Also called flow through distribution.

customer support letter. A letter from a customer or consumer that provides a third-party or independent evaluation of your product.

daily sanitation report. A report generated by an appointed staff member detailing the cleanliness, safety, and suitability of premises for food production.

direct sales. Marketing and selling products, direct to consumers, away from a fixed retail location.

distributor. Entity that buys products or product lines, warehouses them, and resells them to retailers or direct to the end users or customers.

E. coli **O157:H7.** *Escherichia coli* O157:H7 is an enterhemorrhagic strain of the bacterium *Escherichia coli* and a cause of foodborne illness.

EIR (environmental impact report). A study of all the factors influencing the impact that land development or construction would have on the environment, including population, traffic, schools, fire protection, endangered species, archeological artifacts, and community beauty. Many states require that reports be submitted to local governments before the project can be approved.

ELISA (enzyme-linked immunosorbent assay). An immunological technique for accurately measuring the amount of a substance—for example, the presence of an antibiotic in milk.

environmental sampling. One of the most important parts of any risk assessment is the collection of data needed to support exposure and risk calculations. Use of proper sampling methods is essential for ensuring that data are reliable and defensible.

environmental swabs. A ready-to-use, environmental swab system consisting of a 5-inch, rayon-tipped swab containing letheen, a neutralizing buffer, to facilitate recovery of bacteria.

ergonomics. Workplace layout and equipment design intended to maximize productivity by reducing operator fatigue and discomfort.

farmstead cheese. Cheese made on the farm with milk from the farm.

FDA (U.S. Food and Drug Administration). This is an agency within the U.S. Department of Health and Human Services. The FDA is responsible for protecting the public health by assuring that foods are safe, wholesome, sanitary, and properly labeled. The FDA's responsibilities extend to the 50 United States, the District of Columbia, Puerto Rico, Guam, the Virgin Islands, American Samoa, and other U.S. territories and possessions.

fecal coliforms. A fecal coliform (sometimes faecal coliform) is a facultatively anaerobic, rod-shaped, gram-negative, non-sporulation bacterium. Fecal coliforms are capable of growth in the presence of bile salts or similar surface agents, are oxidase negative, and produce acid and gas from lactose within 48 hours at 44° ± 0.5°C.

fictitious business statement. Also commonly known as a "doing business as" or DBA statement, this is a name statement registered with the state. State laws vary in their filing requirements. Some states do a name check first to ensure that no more than one business files under a given name. Other states may require publication of the fictitious name in a local newspaper.

fixed asset. A tangible piece of property that a firm owns and uses in the production of its income. It is held over the long term, and is not expected to be consumed or converted into cash.

fixed cost. Cost that remains unchanged, irrespective of the output level or sales revenue of a firm. Examples of fixed costs are rent, salaries, wages, depreciation, insurance, and interest.

FOB (freight on board). The price invoiced or quoted by a seller that includes all charges up to placing the goods on transportation at the port of departure specified by the buyer. Also called collect freight, freight collect, or freight forward.

food-grade materials. Any material that, when coming into contact with food or the area near food, is unlikely (considering the application and the environment) to contaminate food with harmful materials above the FDA allowed limit.

foodservice. Any entity or business that serves food to consumers.

FRP (fiber-reinforced plastic). Also called fiber-reinforced polymer, this is a composite material made of a polymer matrix reinforced with fibers.

FSIS (Food Safety and Inspection Service). This is an agency of the U.S. Department of Agriculture (USDA). It is the public health agency responsible for ensuring that the nation's commercial supply of meat, poultry, and egg products is safe, wholesome, and correctly labeled and packaged.

FSMA (Food Safety Modernization Act of 2011). This act was signed into law by President Obama on January 4th, 2011. It aims to ensure that the U.S. food supply is safe by shifting the focus of federal regulators from responding to contamination to preventing it. Under the new law, food companies will be required to identify all potential food hazards and prepare and follow written plans to reduce or eliminate these health hazards.

GMP (good manufacturing practice). GMPs are part of a quality system covering the manufacture and testing of pharmaceutical dosage forms or drugs and active pharmaceutical ingredients, diagnostics, foods, pharmaceutical products, and medical devices. GMPs are guidelines that outline the aspects of production and testing that can impact the quality of a product.

gross sales (or gross revenue). A measure of overall sales that is not adjusted for customer discounts or returns. It represents all sales invoices and does not include operating expenses, cost of goods sold, payment of taxes, or any other charge.

HACCP (Hazard Analysis and Critical Control Points). A food-safety process that examines critical points in food production that are most likely to cause foodborne illness.

hauling (milk hauling). In order for the milk marketing system to function and have milk continuously moving through the physical marketing channel to fill consumer demand, farm milk must be picked up from farms frequently and delivered to processing plants. This movement is referred to as milk assembly and involves bulk milk haulers.

heat-exchange chiller. A system whereby fluid (e.g., milk) is chilled by exposing it to a refrigerated environment in which the heat is removed.

HTST (High-Temperature Short-Time pasteurization). Milk is heated to a required minimum temperature of 161.5°F for 15 seconds.

LEED (Leadership in Energy and Environmental Design). An internationally recognized green building certification system, providing third-party verification that a building or community was designed and built using strategies intended to improve performance in metrics such as energy savings, water efficiency, CO_2 emissions reduction, improved indoor environmental quality, and stewardship of resources and sensitivity to their impacts. Developed by the U.S. Green Building Council (USGBC), LEED is intended to provide building owners and operators a concise framework for identifying and implementing practical and measurable green building design, construction, operations, and maintenance solutions.

liability insurance. Part of the general insurance system of risk financing to protect the purchaser (the "insured") from the risks of liabilities imposed by lawsuits and similar claims.

load consolidation. The process of consolidating transportable loads for economic purposes.

logo. A graphic mark or emblem commonly used by commercial enterprises, organizations, or individuals to aid and promote instant public recognition. Logos are either purely graphic (symbols/icons) or are composed of the name of the organization (a logotype or word mark).

lot. A standardized, fixed measure of mass often used to refer to the standard load of a specific product.

make process. The process describing the making of cheese (or other dairy product), including the costs, from raw milk to end product.

margin. The incremental output that results from an increase (or decrease) in an input.

market. Any one of a variety of systems, institutions, or procedures whereby parties engage in exchange. While parties may exchange goods and services by barter, most markets rely on sellers offering their goods or services (including labor) in exchange for money from buyers.

market niche. A niche market is a focused, targetable portion of a market. By definition, then, a business that focuses on a niche market is addressing a need for a product or service that is not being addressed by mainstream providers. The market niche defines the specific product features aimed at satisfying specific market needs, as well as the price range, the production quality, and the demographics that it intends to target.

market penetration. One of the four growth strategies of the Product-Market Growth Matrix. Market penetration occurs when a company enters/penetrates a market with current products.

market segment (market segmentation). A market segment is a subset of a market made up of people or organizations with one or more characteristics that cause them to demand similar products and/or services based on qualities of those products such as price or function.

marketing order. A regulation of an executive agency that sets prices and other conditions for the sale of certain goods. The Agricultural Marketing Service of the USDA uses marketing orders to regulate the sale of dairy products as well as fruits and vegetables.

master distributors. Professional networkers that command a large following and can deliver to anywhere between 500 to 30,000 people, depending on their scope of influence.

master sanitation schedule. An agreement between two or more parties regarding a set of chores that will be completed during a specific time period.

microbe-specific swabbing. Surface areas sampled with the use of a sterile template. A moistened swab is used to rub the exposed surface, after which it is returned to its holder and tested for the presence of microorganisms.

Milk Advisory Boards. Also called Milk Marketing Boards, they are a producer-funded division of a state's Department of Food and Agriculture. They exist in any state that has a milk marketing order, and they execute advertising, public relations, research, and retail and foodservice promotional programs for dairy products.

Milk and Dairy Food Safety Division. States have a milk safety branch, typically within the state agricultural department. It is charged with the mission and responsibility of ensuring that various states' milk, milk products, and products resembling milk products are safe and wholesome, meet microbial and compositional requirements, and are properly labeled. It inspects dairy farms and milk-processing plants, and collects samples of milk and milk products to assure consumer safety. It also ensures that tests used to determine the basis of payment for milk are accurate; and it evaluates dairy farms, milk plants, and laboratories for the U.S. Food and Drug Administration (FDA) for dairy products in interstate commerce.

milk handler. An institution, creamery, or cooperative contracted by a producer to collect, transport, and process the milk from the producer's farm.

milk handler bond. A bond protects the party requesting the bond against any financial losses as a result of poor financial decisions, damages, unethical decisions, or a failure to follow state and local laws on the part of you, the principal. The California Milk Handlers Bond holds you accountable for your business decisions.

milk pooling. The statewide pooling of revenue from dairy processors and the redistribution of those pooled revenues to dairy farmers. This pooling of revenues with regulated distribution to dairy farmers helps ensure that dairy farmers receive an equitable and fair price for the milk they produce. These activities maintain satisfactory marketing conditions and assist in providing stability and prosperity to dairy farmers.

mission statement. A formal, short, written statement of the purpose of a company or organization. The mission statement should guide the actions of the organization, spell out its overall goal, provide a sense of direction, and guide decision making. It provides the framework or context within which the company's strategies are formulated.

MSDS (material safety data sheets). A document that contains information on the potential health effects of exposure to chemicals, or other potentially dangerous substances, and on safe working procedures users should adhere to when handling chemical products.

NCIMS (National Conference on Interstate Milk Shipments). The main function of the Conference is to deliberate proposals submitted by various individuals from state or local regulatory agencies (FDA, USDA, producers, processors, consumers, etc.), who have an interest in ensuring that the dairy products we consume are safe. The proposals are assigned to one of three councils, who then discuss the merits of those particular proposals, with a resulting recommendation to the delegate body.

NEPA (National Environmental Policy Act). This was enacted in 1970 to institute a national policy of environmental protection. The California Environmental Quality Act (CEQA) is a California statute that requires state and local agencies to identify the significant environmental impacts of their actions and to avoid or mitigate those impacts, if feasible. It has become a model for environmental protection laws in other states. Check your state for its equivalent.

organic. Organic foods are those that are produced using environmentally sound methods that do not involve modern synthetic inputs such as pesticides and chemical fertilizers, do not contain genetically modified organisms, and are not processed using irradiation, industrial solvents, or chemical food additives.

organic certification. A certification process for producers of organic food and other organic agricultural products. In general, any business directly involved in food production can be certified, including seed suppliers, farmers, food processors, retailers, and restaurants. Requirements vary from country to country, and they generally involve a set of production standards for growing, storage, processing, packaging, and shipping.

organoleptic. A term related to the senses (taste, sight, smell, touch), it is also a term used to describe traditional USDA meat and poultry inspection techniques, because inspectors perform a variety of such procedures (involving visually examining, feeling, and smelling animal parts) to detect signs of disease or contamination. These inspection techniques alone are not adequate to detect invisible foodborne pathogens that now are the leading causes of food poisoning.

OSHA (Occupational Safety and Health Administration). Congress created OSHA to ensure safe and healthful working conditions for working men and women by setting and enforcing standards and by providing training, outreach, education, and assistance. The OSH Act covers employers and their employees either directly through federal OSHA or through an OSHA-approved state program. State programs must meet or exceed federal OSHA standards for workplace safety and health.

pallet. Sometimes called a skid, this is a flat transport structure that supports goods in a stable fashion while being lifted by a forklift, pallet jack, front loader, or other jacking device.

pasteurize. Pasteurization is a process of heating a food, usually liquid, to a specific temperature for a definite length of time, and then cooling it immediately. This process slows microbial growth in food. Unlike sterilization, pasteurization is not intended to kill all microorganisms in the food. Instead pasteurization aims to reduce the number of viable pathogens so they are unlikely to cause disease. Certain food products, like dairy products, are superheated to ensure that pathogenic microbes are destroyed.

pest management plan. A plan outlining the control and management of pests.

pH. A measure of the acidity or basicity of an aqueous solution. Pure water is said to be neutral, with a pH close to 7.0 at 77°F (25°C). Solutions with a pH less than 7 are said to be acidic, and solutions with a pH greater than 7 are basic or alkaline.

PHSBPRA (Public Health Security and Bioterrorism Preparedness and Response Act of 2002). This act is intended to establish new requirements for registration of possession, use, and transfer of select agents and toxins that could pose a threat to human, animal, and plant safety and health. An important component to the new rules includes security risk assessment of individuals who have access to the select agents and toxins. Any person who meets the criteria of a "restricted person" as defined in the Uniting and Strengthening America by Providing Appropriate Tools Required to Intercept and Obstruct Terrorism Act (USA PATRIOT Act of 2001) must not be allowed to access these materials.

PMO (Pasteurized Milk Ordinance). Regulations approved by the Food and Drug Administration governing the design and maintenance of dairy farms and processing plants to make sanitation and milk quality uniform across state lines.

point-of-sale materials. Promotional material available where a sale takes place.

potable water. Potable water is free from pollution, harmful organisms, and impurities.

price elasticity. In economics, elasticity is the ratio of the percentage change in one variable to the percentage change in another variable. It is a tool for measuring the responsiveness of a function to changes in parameters in a unitless way. Frequently used elasticities include price elasticity of demand, price elasticity of supply, and income elasticity of demand. Elasticity is a popular tool among empiricists because it is independent of units and thus simplifies data analysis.

process control. Maintaining the output parameters (such as pH, salt, or fat content) of a specific process within a desired range.

producer-handlers. Dairy farmers who process milk from their own cows in their own plants and market their packaged fluid milk and other dairy products themselves. Producer-handlers sometimes are referred to as producer-distributors. Producer-handlers may sell products directly to consumers through their own stores, directly to consumers on home-delivery routes, or to wholesale customers such as food stores, vendors, or institutions.

product attributes. When used in a market research context, attributes are simply properties of a given product, brand, service, advertisement, or any object of interest. A product, service, or brand can have many attributes, including cost, value for money, prestige, taste, usability, liking ("affect"), and a wide range of image or personality attributes.

product liability insurance. Product liability is the area of law in which manufacturers, distributors, suppliers, retailers, and others who make products available to the public are held responsible for the injuries those products cause. Although the word *product* has broad connotations, product liability as an area of law is traditionally limited to products in the form of tangible personal property. Product liability insurance is simply insurance acquired for the purposes of reducing the liability for the unintentional injuries caused by a product.

product mix. The composition of goods and services produced and/or sold by a firm. A limited product mix tends to increase the firm's risk at the same time it increases the potential for large profits.

QA/QC (quality assurance/quality control). Quality control emphasizes testing of products to uncover defects, as well as reporting to management, who then makes the decision to allow or deny the product's release. Quality assurance, on the other hand, attempts to improve and stabilize production and associated processes in order to avoid or at least minimize issues that led to the defects in the first place.

random weight. Random-weight items are products sold individually (such as a T-bone steak or a turkey breast) or in packages (such as a multipack of chicken legs or pork chops) that vary in weight. Most meat and produce items sold in supermarkets are random-weight products. In contrast, fixed-weight items (such as packaged lunch meat) are sold in standard-sized packages with fixed weights.

raw materials. A raw material or feedstock is the basic material from which a product is manufactured or made, but it is frequently used with an extended meaning. For example, the term is used to denote material that came from nature and is in an unprocessed or minimally processed state.

raw milk. Milk that has not been pasteurized or homogenized.

recall (food recall plan). A product recall is a request to return to the maker a batch or an entire production run of a product, usually due to the discovery of safety issues.

relative humidity. A measurement of the amount of water vapor in a mixture of air and water vapor. It is most commonly defined as the partial pressure of water vapor in the air-water mixture, given as a percentage of the saturated vapor pressure under those conditions. The relative humidity of air thus changes not only with respect to the absolute humidity (moisture content) but also to temperature and pressure, upon which the saturated vapor pressure depends. Relative humidity is often used instead of absolute humidity in situations where the rate of water evaporation is important, as it takes into account the variation in saturated vapor pressure.

Retail Food Facility Law. States have laws regulating retail food codes in order to establish statewide standards for sanitation and health in retail food facilities. In California, it is called the Uniform Retail Food Facilities Law (CURFFL).

RFID (radio-frequency identification). A technology that uses communication via radio waves to exchange data between a reader and an electronic tag attached to an object, for the purpose of identification and tracking. RFID makes it possible to give each product in a grocery store its own unique identifying number. It can provide nearly anything (assets, people, work in process, medical devices, etc.) with its individual unique identifiers—like the license plate on a car, but for every item in the world. RFID has many applications; for example, it is used in enterprise supply chain management to improve the efficiency of inventory tracking and management.

sanitize. To make sanitary, as by cleaning or disinfecting.

scale-up. To proportionally increase a process from the lab scale to the pilot plant scale or commercial scale.

shelf talkers. Also called point-of-purchase display, this refers to display of product and/or product information at the place of purchase.

slotting fee(s). See allowances.

specialty cheese. A value-added product that commands a premium price. According to the Wisconsin Specialty Cheese Institute, the nature of specialty cheese is derived from one or more unique qualities, such as exotic origin, particular processing or design, limited supply, unusual application or use, and extraordinary packaging or channel of sale. The common denominator is its very high quality.

SSOP (sanitation standard operating procedure). This is the common name given to the set of sanitation procedures in food production plants that are required by the Food Safety and Inspection Service of the USDA and regulated by 9 CFR Part 416 in conjunction with 21 CFR Part 178.1010. It is considered one of the prerequisite programs of HACCP.

standards of identity. In the context of food, these are the mandatory, federally set requirements that determine what a food product must contain to be marketed under a certain name in interstate commerce. Mandatory standards (which differ from voluntary grades and standards applied to agricultural commodities) protect the consumer by ensuring that a label accurately reflects what is inside (for example, that mayonnaise is not an imitation spread, or that ice cream is not a similar, but different, frozen dessert). They are issued by the USDA, the FDA, or the Bureau of Alcohol, Tobacco, Firearms and Explosives.

sterilizing. To make free from live bacteria or other microorganisms.

stop charge. Fees charged by milk hauler based on distance from plant, ease of access, etc.

succession plan. A process for identifying and developing internal people with the potential to fill key leadership positions in the company.

target market. A group of customers at which the business has decided to aim its marketing efforts and its merchandise. A well-defined target market is the first element of a marketing strategy.

total cost. In economics and cost accounting, total cost (TC) describes the total economic cost of production. It is made up of variable costs (which vary according to the quantity of a good produced and include inputs such as labor and raw materials) and fixed costs (which are independent of the quantity of a good produced and include inputs–capital–that cannot be varied in the short term, such as buildings and machinery). Total cost in economics includes the total opportunity cost of each factor of production as part of its fixed or variable costs.

U.S. Green Building Council. The U.S. Green Building Council is a 501(c)(3) nonprofit trade organization that promotes sustainability in how buildings are designed, built, and operated.

variable costs. Periodic costs that vary with the output or the sales revenue of a firm. These include raw materials, energy usage, and labor.

variance (zoning variance). A variance is a requested deviation from the set of rules a municipality applies to land use known as a zoning ordinance, building code, or municipal code. In today's land use approval climate, a variance request can be a fatal flaw for something as "insignificant" as a variation from a municipal sign ordinance.

vertical integration. This is when all stages of production are controlled by one company, from the acquisition of raw materials to the retailing of the final product.

water activity (a_w). The water activity of a food is the ratio between the vapor pressure of the food itself, when in a completely undisturbed balance with the surrounding air media, and the vapor pressure of distilled water under identical conditions. A water activity of 0.80 means the vapor pressure is 80 percent of that of pure water. The water activity increases with temperature. The moisture condition of a product can be measured as the equilibrium relative humidity (ERH) expressed in percentage or as the water activity expressed as a decimal. Most foods have a water activity above 0.95, and that will provide sufficient moisture to support the growth of bacteria, yeasts, and mold. The amount of available moisture can be reduced to a point that will inhibit the growth of the organisms. If the water activity of food is controlled to 0.85 or less in the finished product, it is not subject to the regulations of 21 CFR Parts 108, 113, and 114.

MEASUREMENT CONVERSION TABLE

U.S. Customary	Conversion factor for English to metric	Conversion factor for metric to English	Metric
inch (in)	2.54	0.394	centimeter (cm)
foot (ft)	0.3048	3.28	meter (m)
square foot (ft²)	0.0929	10.764	square meter (m²)
acre (ac)	0.4047	2.47	hectare (ha)
gallon (gal)	3.785	0.26	liter (l)
ounce (oz)	28.35	0.035	gram (g)
pound (lb)	0.454	2.205	kilogram (kg)
Fahrenheit (°F)	°C = (°F − 32) ÷ 1.8	°F = (°C × 1.8) + 32	Celsius (°C)

INDEX ∽

Page numbers in **bold type** indicate major discussions. Page numbers in *italic type* indicate tables.

A

accessibility standards, 123
access roads, 10
ACS (American Cheese Society), 2, 7, 47, 70, 72
ADA (Americans with Disabilities Act), 123
advertising, 69–70, 73
aging of cheeses, producer survey, *67*
aging rooms, 40–41
aging times, various cheeses, *41*
allergens, protocol for, 101–103
AMA (American Marketing Association), 66
American Cheese Society (ACS), 2, 7, 47, 70, 72
American Marketing Association (AMA), 66
Americans with Disabilities Act (ADA), 123
artisan cheese, 2
Asiago cheese, *41*
ATP (adenosine triphosphate) detection, 95, 98, 100
audits, 80, *81*, 103

B

banks, 14–16
barcodes, 57, *81*
batch pasteurization, 39
benefits of cheesemaking business, 7
blogs, 72
brands, 64, 66, 72
Brie cheese, *41*
brining rooms, 40
broker sales, 46–47
budgets. *See* finances
buildings conversion, 14, 123
business license, 124
business name, recording of, 124
business plans, **15–28**
    concept descriptions, 18, 20, 22
    goal setting, 20–23
    importance of, 15–16
    mission statements, 16–19
    outline, 28
    product identification, 17
    resources for, 16, 25–27
    sequence of tasks, 24
    target market identification, 17

C

California Artisan Cheese Guild, 7
California Department of Food and Agriculture (CDFA)
    cheese plant inspections, 127
    disease outbreak notification, 105
    milk production permits/milk prices, 124, 125
    organic registration, 124
    pasteurizer equipment certification, 126
    recall decisions, 106
    summary of responsibilities, *128*
California Food and Agriculture Code, 126
California Milk Advisory Board (CMAB), 2, 3
California Travel and Tourism Commission, 71
California Uniform Retail Food Facilities Law (CURFFL), 124
CALTRANS, *128*
capital, 11, 14, 16
    See also finances
case cards, 51, 59, *61*
    See also shelf talkers
CDFA. *See* California Department of Food and Agriculture
Centers for Disease Control (CDC), 105
Certified Producer's Certificate (CPC), 124
chambers of commerce, 26
charge backs, 48
Cheddar cheese, *41*, 113
cheese buyers, 49, 51, *61*
cheese plants. *See* plant layout and design
cheeses. *See individual cheeses:* Mozzarella, goat, etc.
cheese vats, 34
chemicals, storage and use, 94, 100–101
Chèvre, 114, *115*
cleaning. *See* sanitation
cleaning products, storage and use, 94, 100–101, *102*
clean in place (CIP), 89, 100
clean out of place (COP), 89, 100
clothing procedures, for food safety, 88–89, 91
CMAB. *See* California Milk Advisory Board
coliform bacteria, 90
collateral, 11, 14
    See also finances
competition, identifying, 7, 51, 57
construction materials, 32–33
consumers. *See* customers
consumption statistics, 3, 57, 60
contamination. *See* food-safety strategies; *Listeria* spp.;
    microbial growth
contingency planning, 117–118
co-packing, 24, 66
Costco, 51, 116
costs of cheesemaking business, 7
    See also finances
Cotija cheese, 114, *115*
Cowgirl Creamery, 72
CPC (Certified Producer's Certificate), 124
cross-docking, 53
CURFFL (California Uniform Retail Food Facilities Law), 124
customer complaint form, 104

customers
    buying practices of, *56*, 58–59, 68
    complaints from, 103–105, 125
    identifying, 7–8, 17,
    perception of brands, 66
    surveys of, 58–59, 60, 61, 67, 71
    understanding, 58–61
    customer service, 73, 103–105
    customer support letters, 14, 24

D

dairy inspector, working with, 32, 33
Dean and Deluca, 2, 51
debt financing, 15
    See also finances
Delicious, Web site, 71
demographics, buyer, 56
    See also customers, surveys of
demos, in-store, 48–49, 51, 70
Department of Health Services, *128*
Department of Industrial Relations, *128*
design. *See* plant layout and design
Digg, social news Web site, 71
direct marketing, *70*
direct sales, 21–23, 46, *70,* 72, 117, 124
discounts
    customer response to, 67
    for inventory clearance, 70
    on large-volume sales, 51, 52, 117
    sellers' allowances and promotions, 48, 49, 69
disease outbreak, 105
    See also product recall
disinfectants, *101, 102*
distribution logistics, 54, 85
distribution systems, **46–54**
    brokers, 46–47
    direct sales, 21–23, 46, 70, 72, 117
    distributors, 47–49, 50, 52, 61
    foodservice, 52, 61
    retail, 49, 51, 61, 69, 116, 123
distributor sales, 47–49, 50, 52, 61
documentation, 80, *81,* 96–97, 102, 104–105, 110–111
drains, type and placement, 33

E

EIR (environmental impact report), 123
electricity needs/costs, 39–40
electronic media. *See* social media
ELISA (enzyme-linked immunosorbent assay) test kits, 103
Emmental cheese, *41*
employees
    changing areas, restrooms for, 34–35, 93
    ergonomics and plant design, 33–34, 36
    as goodwill ambassadors, 73
    health and hygiene, *81,* 84, 93, 94–95

    training. *See* training of employees
    work areas, separated, 34
    See also food-safety strategies
energy cost control, 36, 38–40
environmental controls, heating, cooling, etc., 36, 38–40, 84
Environmental Health Department, 105, 123–124, 125, *130,* 131
environmental impact report (EIR), 123
environmental swabbing. *See* swabbing procedures
equipment for cheesemaking, 33–42
    cleaning, 33, 34, 80, *81*
    ergonomics, 33–34, 36
    good management practices, 83–85
    See also sanitation
equity financing, 15
ergonomics, 33–34, 36
European cheeses, 3, 41, 51
evaluation processes, **6–11**
    business skills, 7, 9
    competition, 7, 57
    family commitment and fit, 6–7, 16–17, 18
    Four P's marketing tools, 63–73
    marketplace and consumer trends, 56–58
    physical resources, 10–11
    price setting, 66–69
    risk management, 116–117
    sanitation procedures, 95–98, 100
    targeting the market, 62
    See also business plans; finances
executive summary, part of business plan, 16
exit strategy, 25
expansion of facilities, 32, 41–42, 73, 116

F

Facebook, 71, 72, 73
facility design and planning, 14, 123
    See also plant layout and design
family involvement, 6–7, 16–17, 18,
Fancy Food Show, 7, 47, 70
Farm Credit Service, 15
Farm Credit System, 26
farmers' markets
    costs of, 46
    customer research, 7
    demos at, 49
    insurance, 114
    permits, 124
    promotion tool, *70*
Farm Service Agency, 15, 27
farmstead cheese, 2
farm store, 10
FDA. *See* Food and Drug Administration
Feta cheese, 2
fiberglass-reinforced plastic (FRP) panels, 32
finance plan/projections, part of business plan, 23–24, 28

finances
    of adding a new enterprise, 6–7
    economic trends, 56
    financial statements, 23–24
    loan sources, 11, 14–16
    pricing strategies, 66–69
    resources about, 15, 23–24, 26–27
    risk management, 116–118
    start-up costs, 11, 14–15, 24–25
    See also goal setting; margins; pricing
FOB (freight on board), 47–48, *53*
Food and Drug Administration (FDA),
    defect levels, rules for, 83–84
    employee training requirements, 83
    food recall involvement, 106
    food-safety survey, 80, *81*
    good management practices (GMPs), 83–85, 86, 131
    labeling compliance guidelines, 103, 127
    Pasteurized Milk Ordinance (PMO), 126, 129
    registration with, 129
    summary of responsibilities, *128, 130*
foodborne disease outbreak, 105
    See also product recall
foodborne illness, 80, *81, 99,* 105, 112
    See also sanitation
Food, Drug, and Cosmetic Act, Title 21, Part 110, 83–85
food handler's certificate, 124
food pairings, 59
Food Safety Inspection Service, 94, 106
Food Safety Modernization Act, 105, 129, 131
food-safety strategies
    overview, 80–82
    allergens, protocol for, 101–103
    good management practices (GMPs), 83–85, 86, 131
    grocery chain requirements, 51
    liability insurance, 51, 114
    process control protocols, 112–114
    product sampling, 98, 99
    recall procedures, 105–111
    sanitation master plan, 85–97
    top problems, 80, *81*
    transportation considerations, 48
    See also plant layout and design; regulations; SSOPs
foodservice sales, 52, 61
footbaths, 34, 83, 92
Four Ps, tools for marketing plan, 63–73
freight, 47–48, 49–50, 51, 53–54, 85
freight on board (FOB), 47–48, 53
FRP (fiberglass-reinforced plastic) panels, 32

G
garbage disposal, 85
gas needs/costs, 39–40
GMPs (good manufacturing practices), 32, 33, 34, 83–85, 86, 131

goal setting, 20–23, 62–63
goat cheese, sales growth, 2
goat's milk, 124
good manufacturing practices (GMPs), 32, 33, 34, 83–85, 86, 131
Gouda cheese, *41*
government agencies. *See* California Department of Food and Agriculture; Food and Drug Administration; regulations
grocery stores. *See* retail sales
growth in industry, 2–3
growth, planning for, 41–42, 62–63, 64, *73,* 116–117
guilds, 7, 72

H
HACCP (Hazard Analysis and Critical Control Point) certification, 51, 86, 105, 114
Health Department, county, 105, 123–124, 125, *130,* 131
health, of personnel, 84, 94–95
heat requirements, calculating, 39
high-temperature short-time (HTST) pasteurization equipment, 39
history of U.S. cheesemaking, 2
hoses, location and connections, 33, 85, 90
HTST (high-temperature short-time) pasteurization equipment, 39
humidity needs, for maturation, *41*
hygiene, of personnel, *81,* 83, 84, 85, 93

I
infrastructure assessment, 10
inspections
    importance of, 80
    milk production, 125, 127
    pest management, 95, 98
    plant design, 32, 33, 126
    regulatory agencies, *128, 130,* 131
    sanitation reports, 96, 97
    See also SSOPs
Internet marketing. *See* social media
Internet sales, 46, 54, 57

L
labeling
    allergenic ingredients, 101–103
    brands, 64, 66
    consumer survey, 59
    customer service information, 73
    incorrect, safety risk, *81, 106*
    regulations about, 127
    retail chain requirements, 51
    See also shelf talkers
labor. *See* employees
land resources and use, 10, 14, 122–123
Laura Chenel, 25

layout. *See* plant layout and design

lease negotiation, 116

LEED (Leadership in Energy and Environmental Design) certification, 39

legal constraints. *See* regulations

liability insurance, 51, 114

lines of credit, 15

Listeria spp., 33, 98

load consolidation, 53

loan sources, 14–15

low-fat cheeses, 59

low-sodium cheeses, 59

M

mail-order sales, 54, 72

margins

    calculation example, 53

    distributor sales, 49, 53

    foodservice, 52

    retail sales, 51, 53

marketing channels. *See* distribution systems

marketing narrative information, 21, 48, 51, 59, *61*, 64

marketing strategies, **54–77**

    overview, 54–55

    brands, 66

    customer surveys, *56*, 58–59, 60, 61, 67, 71

    Four P's, 63–73

    for new products, 48, 55, 63, 69–70

    niche marketing, 7, 61–62, 66, 72

    objectives setting, 62–63

    pricing, 66–69

    product, components of, 64–66

    promotion, 69–73

    SWOT (strengths, weaknesses, opportunities, and threats), 56, 57–58

    target market, focus on, 57–58, 62

    tracking progress, 73–77

    trend analysis, 56–57

markets, diversity of, 116

master distributor, 52

material safety data sheet (MSDS), 101

maturation of cheeses, 40–41

microbial growth, 85, 98, *99*, 101, 112–114, *115*

    See also testing, for contamination

milk

    marketing orders, 124

    regulations, 124–126, 127

Milk and Dairy Food Safety Branch, 32, 124, 127, 128

mission statement, part of business plan, 16–19, 62

moisture content, and food safety, 112–114, *115*

monitoring procedures. *See* documentation; SSOPs

Mozzarella, 69, 113

MSDS (material safety data sheet), 101

multiple-channel distribution, 49

N

NAFDMA (North American Farmers' Direct Marketing Association), 72

name of business, recording, 124

narrative information, as sales tool, 21, 48, 51, 59, *61*, 64

niche, product, 7, 17, 61–62, 66, 72

NOP (National Organic Program), 124

North American Farmers' Direct Marketing Association (NAFDMA), 72

O

objectives, in business and marketing plans, 20–23, 62–63

Occupational Safety and Health Administration (OSHA), 101, *128*

online reviews, 71–72

online store, 46, 54, 57, 72

organic production, 91, 124, 129

OSHA (Occupational Safety and Health Administration), 101, *128*

P

packaging, 51, 64, *67, 81, 106*

Parmesan cheese, *41*

pasteurization, 39, *81*, 126, 127, 129, 130

pathogen contamination, 33, 98, *99, 106*

    See also microbial growth

payment terms, 46, 48, 64

Pedrozo market survey, 60

permits for

    summary, 130

    buildings, 32–33

    chemical applications, 98

    farmers' market booths, 124

    highway signs, *128*

    land use, 122–123

    milk production, processing, hauling, 124–125, *128*

    operation of cheesemaking business, 127

    septic systems, 123

    wastewater discharge, 123–124, 128

pest prevention and control, 33, 34, 84, 95, 98, 100

pets, and food safety, 100

pH, and food safety, 112–114, *115*

planning. *See* business plans; evaluation processes; expansion of facilities

plant layout and design, **32–42**

    bathrooms, 36, 93

    brining and aging rooms, 40–41

    cleaning, 32, 33, 34, 40, 80, *81*

    construction materials, 32–33

    energy savings, 36, 38–40

    ergonomics, 33–34, 36

    expansion, 32, 41–42

    floor plans, 35–37

    good management practices (GMPs) and, 83

people flow, 34–37

pest control, 33, 34, 84, 95, 98, 100

processing flow, 34, 38, 42

product flow, 34–38

regulations about, 84–85, 126–127

plumbing, design, 33, 85

point-of-sale materials, 21, 48–49, 51

See also case cards; shelf talkers

positioning, part of marketing, 62

press and public relations, 70

price breakpoints, 68

pricing, 51, 52, 53, 59, 66–69

See also distribution systems

processing flow diagram, 38

product contamination control. See food-safety strategies

product diversity, risk management, 116

product, expanded definition of, 64

product flow diagram, 38

production statistics, 2

product liability insurance, 51, 114

product mix, 64

product quality

aging room environment, 40–41

customer perceptions of, 56, 66, 68

grocery chain requirements, 51

process control (pH, salt, moisture), 112–114, 115

See also food-safety strategies; product recall

product recall, 105–111

classifications, 106

communications, 105–106, 107–109

customer notification log, 110

examples, 106

government agencies involved, 105, 106

notification letter, sample, 111

product identification sheet, 111

recall plan, sample, 107–110

product research and segmentation, 7, 61–62

product sampling, risk management, 98, 99

profits. See pricing

promotion, 69–73

advertising, 69–70, 73

distributor sales, 48–49

monitoring effectiveness, 73

social media, 71–73

tools and their use, 70

Public Health Security and Bioterrorism Preparedness and Response Act, 129

publicity, 70

See also promotion

public relations, 70

See also promotion

quality. See product quality

quick response (QR) code, 51

QR (quick response) code, 51

radio-frequency identification (RFID), 57

recall of product. See product recall

recordkeeping. See SSOPs

Reddit, reviews Web site, 71

Regional Water Quality Control Board, 123, 124, 128

regulations, 122–131

agency responsibilities summarized, 122, 128

building codes, 123

business license and taxes, 124

farmers' market permits, 124

Food, Drug, and Cosmetic Act, Title 21, Part 110, 83–85

good management practices (GMPs), 83–85, 86, 131

land use, 122–123

milk production, processing, and sales, 124–126, 127, 129

organic registration and certification, 124, 129

OSHA (Occupational Safety and Health Administration), 101, 128

plant design and operation, 126–127, 131

product identity and labeling, 103, 127, 129

public health and security, 123–124, 129, 131

resources about, 85, 94, 126, 127, 129

weights and measures, 124

resources for

business plans, 16, 25–27

finances, 15, 23–24, 26–27

process control, 113

regulations, 85, 94, 126, 127, 129

sanitation, 85, 94

restaurant sales. See foodservice sales

retail sales, 49, 51, 61, 69, 116

See also pricing

revenue. See pricing

RFID (radio-frequency identification), 57

risk management plans. See food-safety strategies; finances, risk management

Roquefort cheese, 41

rubbish disposal, 85

rural economic development agencies, 15

safety. See finances, food-safety strategies; risk management

Safeway, 51

sales promotion. See promotion

sales statistics, 2, 3

salt concentrations, 112–114, 115

samples, 51, 59, 70

sanitation, **83–98**
    for allergenic products, 102–103
    FDA study, a top problem, *81*
    good management practices (GMPs), 83–85, 86, 131
    master plan, 85–98
    regulations, 83–85, 123–124
    report and scheduling forms, 96, 97
    resources about, 85, 94
    swabbing procedures, 95, 98, 100
    See also food-safety plans; plant layout and design;
      SSOPs
sanitation standard operating procedures. *See* SSOPs
sanitizers, 91, 94, 101
SBA (Small Business Administration) 15, 16, 26
SBCDs (Small Business Development Centers) 16, 26
scaling up, 32, 41–42
    See also growth, planning for
septic system permit, 123
sheep's milk, 124
shelf talkers, 51, 60–61, 73
shipping. *See* distribution systems; freight
sinks, placement and design, 33
Small Business Administration (SBA), 15, 16, 26
Small Business Development Centers (SBDCs), 16, 26
social media, 71–73
SSOPs (sanitation standard operating procedures), 85–98
    cleaning sample SSOPs, 88–90, 91–92
    cross-contamination sample SSOPs, 92–93, 94
    daily/weekly report forms, 96, 97
    employee health sample SSOP, 94–95
    pest management sample SSOP, 95
    water quality sample SSOPs, 87–88, 90–91
staffing. *See* employees
start-ups
    financial strategies, 11, 14–16, 24–25, 67
    land and buildings, 14, 123
    marketing plans, 64, 73
State Organic Program, 124
statistics, 2–3
sterilizers, 101
stories, as sales tool, 21, 48, 51, 59, *61,* 64
Straus Family Creamery, 72
Stumble Upon, Internet community, 71
succession, business, 25
swabbing procedures, 95, 98, 100
SWOT (strengths, weaknesses, opportunities, and
    threats) analysis, 56, 57–58

T
target markets, 57–58, 62
temperature needs, for maturation, *41*
terroir, assessment of, 10, 64
testing, for contamination, 95, 98, 100, 103
Title 21, Part 110, Food, Drug and Cosmetic Act, 83
trade publications, 7, 70

Trader Joe's, 116
training of employees
    overview, 83
    allergen procedures, 103
    chemical use, 101
    customer complaints, 105
    FDA study, 80, *81*
    good management practices, requirements, 84
    importance of, 80
transportation. *See* access roads; freight
TripAdvisor Media Network, 71, 72
Twitter, 71, 72, 73

U
University of Guelph, 40–41
U.S. Department of Agriculture (USDA), 3, 15, 27, 103,
    128, 129
use permits. *See* land resources and use
U.S. Food Safety and Inspection Service, 94, 106

V
vendor approval, 24, 49
vertical integration, 2–3, 6, 20
visitors, attracting, 10

W
waste disposal, 84, 85
water activity, and microbial growth, 112–114, *115*
water cost savings, 38, 39
water supply, design and quality 33, 85, 87–88, 90–91, 123
Web sites, as marketing tool, 57, 71–73
weights and measures department, 124
whey disposal, 57, 123–124
Whole Foods, 2, 62
worksheets
    customer base, 8
    business concept, 20
    business skills, 9
    competitive advantage, 76, 77
    competitors, identifying and ranking, 75, 76
    Four P's marketing tools, 65
    mission statement, 19
    permit and inspection costs, 130
    physical resources, 10
    priorities, business and family, 18
    product assessment, 74
    SWOT analysis, 58

Y
YouTube, 71

Z
zoning ordinances, 122–123